ETHICS FOR ENVIRONMENTAL POLICY

THIRD EDITION

By Albert Berry Crawford

cognella
SAN DIEGO

Bassim Hamadeh, CEO and Publisher
Carrie Montoya, Manager, Revisions and Author Care
Kaela Martin, Project Editor
Abbey Hastings, Associate Production Editor
Jess Estrella, Senior Graphic Designer
Alexa Lucido, Licensing Manager
Natalie Piccotti, Director of Marketing
Kassie Graves, Vice President of Editorial
Jamie Giganti, Director of Academic Publishing

Printed in the United States of America.

cognella® | ACADEMIC PUBLISHING
3970 Sorrento Valley Blvd., Ste. 500, San Diego, CA 92121

This book is dedicated by my wife, Leticia, whose love of gardening and efforts to preserve the monarch butterfly are inspirational; to our son, Berry, whose imagination is expressed in diverse and surprising ways; to our son, Charles, for whom nature is a source of joy and contentment; and to our daughter, Georgina, whose veterinary care for animals is practiced with skill and compassion.

BRIEF CONTENTS

DETAILED CONTENTS

INTRODUCTION

This third edition of *Ethics for Environmental Policy* contains two new chapters (1 and 2), together with a substantial reorganization of materials presented in the previous two editions.

Chapter 1 ("A Short Primer on Ethics") provides an overview of ethical concepts and theory, with emphasis on how various theoretical orientations have been applied in debate in animal and environmental ethics and how these debates have, in turn, shaped practice and policy. The overview of ethical theory discusses three main areas of ethical inquiry (conduct ethics, virtue ethics, and metaethics), highlights a number of decision principles (including formulations of the principles of self-interest, utility, nonmaleficence, equal worth, equal consideration, precaution, and sustainability), analyzes the concept of moral standing (among other central ethical concepts), and reviews recent work in environmental virtue ethics.

Chapter 2 ("Environmental and Ecological Decline") presents an extended discussion of three interrelated global problems: climate change, biodiversity loss, and the huge quantities of domestic and industrial wastes released to the planet's atmosphere, land, and water. Reports of international and national assessments are reviewed, including the 2013 and 2018 Reports of the UN's Intergovernmental Panel on Climate Change (IPCC), the 2015 Paris Agreement on Climate Change, and the UN's 2019 Intergovernmental Science-Policy Platform on Biodiversity and Ecosystem Services. The purpose of this new chapter is to highlight the seriousness of these ominous trends, the practical relevance of issues addressed in academic ethics, and the need for more focused ethical commentary in policy debates and institutional responses.

Chapter 3 ("Views of Nature") discusses the many ways in which the relationship of humans to the natural order have been conceptualized in various religious and cultural traditions. The chapter concludes with comments on how the concept of nature is being reconceptualized, influenced largely by developments in ecology and conservation biology.

Chapter 4 ("Contributing Elements to an Attitude of Respect") attempts to distill what many nature and environmental writers emphasize is worthy of respect in nature. This review is organized under the following headings: "Rekindling Felt Connections"; "The Primary of Place, Aesthetic Experience: Awe, Beauty, and Wonder"; "Understanding Complex Systems"; "Telling Life's Story"; and "Taking Stock of Nature's Manifold Value." The chapter concludes with an analysis of four primary environmental virtues (respect, humility, gratitude, and responsibility) and how, together, these virtuous dispositions constitute the core of what it means to be "rightfully oriented toward nature."

Chapters 5, 6, 7, and 8 examine theoretical perspectives developed by leading writers within each of several orientations ("isms") in contemporary animal and environmental ethics: anthropocentrism, vertebracentrism, biocentrism, and ecocentrism. Various positions that have been staked out on the question of the moral standing of nonhuman biotic entities are discussed as part of this discussion. The systematic ethical theories of animal rights advocate David DeGrazia, biocentrist Paul Taylor, and ecocentrist Aldo Leopold are examined in some detail. The last section of Chapter 8 reviews how the "intrinsic" vs. "instrumentalist" views of nature have played out in major policy debates and how efforts are being made in conservation biology to bring the two camps together.

Chapter 9 ("Capitalism, Markets, and Government") addresses arguments regarding the respective roles that markets and government play, or ought to play, in resource management decision making. To provide context for this discussion, the chapter begins with a discussion of capitalism: it's ideal as formulated in free market economic theory and how its assumptions and "real world" practices are challenged and critiqued. Attention is given to the role played by cost benefit analysis in resource management decision making and to policy initiatives aimed at mobilizing market forces in support of environmental objectives. Cap-and-trade systems for limiting carbon and other emissions are reviewed in this context.

Chapter 10 ("Sustainability, Ecological Footprints, and Future Generations") focuses on three closely related and widely discussed concepts and approaches in environmental ethics. Various definitions of sustainability are linked to obligations owed to future generations, measures of sustainability are typically framed in terms of the size of ecological footprints, and serious concern is mounting that ecological footprints exceed the capacity of the planet to sustain current levels of production and consumption (measured by footprint analysis). Attention is then given to calls for the exercise of precaution in policy making. In this context, various definitions and interpretations of the principle of precaution by United Nations organizations and the European Union are reviewed. Chapter 10 concludes with a brief discussion of the bioregional movement in which many of the ideas about sustainable living are incorporated.

Chapter 11 ("Environmental Impact Assessment: A Modeling Approach") outlines a modeling approach that incorporates multiple evaluation criteria to evaluate how policies and large-scale projects affect both humans and nonhuman biotic entities, both negatively and positively. The model's conceptual structure incorporates the following: (1) designation of aggregate biotic entities potentially affected by a policy or project (e.g., humans, vertebrate animals, invertebrate animals, plant life, imperiled species and/or ecosystems; (2) assembly of predicted effects of the policy or project for each aggregate entity type; (3) characterization of each predicted effect as harmful or beneficial using six designated harm types (physical damage, resource deprivation, functional impairment, physical pain, emotional distress, and life termination) and six benefit types stated as complements (opposites) of the six harm types; and (4) estimation of the probability and degree of magnitude of each predicted harmful or beneficial effect using scalars to estimate the extent, onset, duration, proliferation, controllability, and remediability of each predicted effect. Once the estimated probability and degree-of-magnitude scalers are entered (as inputs) into the appropriate cells of the model's impact matrix, the model computes a number of policy-relevant indices, including total harm and benefit indices at the entity level and a composite benefit/harm ratio for the policy or project evaluated. If the ratio is not greater than 1.0, the policy or project is preliminarily judged not to be morally permissible. If the ratio is greater than 1.0, the threshold test of moral permissibility is met, leaving open the question of whether other conditions of moral permissibility are satisfied. Suggestions are made for the use of the model in a classroom environment using a well-developed case study. Consideration is given to how the model might be used to incorporate value weights in a way that minimizes morally objectionable interspecies bias, and to how the model might be used more generally in policy-making forums. The book's appendix provides a case (for classroom use) pertaining to proposals to open the Alaska Native Wildlife Refuge to oil and gas development.

A few comments about the distinguishing features of the book and reasons why it might be an attractive adoption in college courses may be helpful and not too presumptuous. Because most of the issues in contemporary environmental ethics are included in the book's coverage, it might be used as a text in the field, perhaps supplemented by materials associated with topics that instructors may choose to give emphasis. For courses that have an applied focus and utilize case studies, the book's emphasis on application methodology may make it an attractive choice. Further, because the book incorporates accounts of developments in various sciences and devotes attention to the description of social movements shaping public attitudes and policy, the book may be an attractive supplement in conservation biology or ecology courses that emphasize policy issues and their relation to ethics. It may also be a good choice for environmental studies courses having an issues-related and/or historical orientation.

1 A SHORT PRIMER ON ETHICS

This introductory chapter describes the main branches of ethical inquiry (conduct ethics, virtue ethics, metaethics), identifies and explains central ethical concepts (what constitutes moral rightness, the fundamentals of well-being, the basis of fairness, rights entitlements, etc.), discuses a number of moral principles developed within various theoretical frameworks in ethics, and examines how these various principles are applied in moral deliberation, justification, and critique.

Emphasis is given to the definition of ethical terminology and how ethical concepts and principles have been applied to problems and issues in environmental and animal ethics from anthropocentric, vertebracentric, biocentric, and ecocentric (ecological) theoretical perspectives.

There is, obviously, considerable practical disagreement about how humans ought to treat one another, other biotic entities, and the broader natural environment, much of this disagreement stemming from underlying theoretical disagreement.

AREAS OF ETHICAL INQUIRY

Ethics texts often distinguish three areas of ethical inquiry, typically labeled "conduct ethics," "virtue ethics, and "metaethics."

Conduct ethics focuses on morally right *conduct*; on what it means to *do* the morally right thing; on the formulation and use of moral *principles* to inform morally right decision making (moral deliberation), to justify the choices one makes and acts upon (moral justification), and to evaluate and critique policies, practices, and institutional arrangements (moral critique). Such inquiry is substantive, prescriptive, and centered on what *ought* to be the case.

Virtue ethics focuses on human *ideals*; on *dispositions* to think and act in morally right ways; on defining the elements of *good character*; on the articulation of key moral *virtues*. Here the emphasis is not on how one ought to choose and act, but on how one ought to *be*; on what it fundamentally means to be a morally good *person*. Similar to conduct ethics, virtue ethics is substantive and prescriptive.

Metaethics, in contrast, focuses on the meaning of ethical terms; on conceptual clarification; on classifying and making distinctions between various approaches in conduct and virtue ethics; on analyzing the logical adequacy of moral arguments; on test cases to draw out the implications, limitations, or problematic aspects of prescriptive moral constructs. Metaethics is not itself prescriptive, but descriptive and/or analytical.

It is noted that these characterizations of conduct ethics and virtue ethics are made from the perspective of a **personal ethic** focusing on individual choice and action. In the context of a **social ethic** (or organizational ethic), conduct ethics is oriented toward assessments of the moral rightness of policies and procedures adopted by corporations and governmental bodies to guide organizational decisions, while virtue ethics is oriented toward the articulation of corporate values, mission statements, vision statements, and the like, and toward fostering an organizational culture that reflects organizational values and objectives.

An aphorism in ethics states that "dispositions without principles are impotent; dispositions without principles are blind." One can discourse at length about how one ought to go about deciding what is morally right in particular situations, but if one lacks the virtues that dispose one to think and act in the ways prescribed, the talk is often little more than that: ineffectual talk. On the other hand, being inclined or disposed to think and act in certain ways is also not enough, particularly in the many situations in which interests and obligations conflict.

Conduct ethics and virtue ethics are interdependent and mutually reinforcing, and emphasizing one at the neglect of the other tends to result in moral "impotence" or "blindness."

ETHICAL CONCEPTS

Five basic ethical concepts are discussed next: rightness, well-being (goodness), rights entitlement, justice (fairness), and moral standing.

Moral Rightness: Doing the Right Thing and Being Good[1]

Starting with the concept of moral rightness, the meaning of a number of frequently used terms and distinctions need to be clarified:

- **Moral rightness and moral permissibility.** A *morally right* act is an act that is either *morally obligatory* or *morally permissible*, in which a morally obligatory act is defined as an act that is morally required (enjoined), and a morally permissible act is defined as an act that is not morally prohibited.
- **Acts of omission and commission.** This distinction refers to two types of moral wrongness. Failure to do what is morally right (i.e., what one ought to do, or what is morally required) is said to be a wrongful act of *omission*. Doing what is morally wrong (i.e. what one ought not to do, or what is morally prohibited) is said to be a wrongful act of *commission*.
- **Prima facie obligation and actual obligation.** *Prima facie obligation* refers to what one, in general, ought to do and *actual obligation* refers to an obligation judged (in moral deliberation) to take precedence over another obligation in a particular situation. In a situation in which telling the truth would be devastating to a friend and not revealing the truth would not have not have a significant bearing on the friend's well-being, the general (prima facie) obligation to be kind would be judged to take precedence over the general (prima facie) obligation to tell the truth.

A distinction is often made between **teleological** and **deontological** theories of moral obligation. On a *teleological* (or consequentialist) approach, the moral rightness and wrongness of an action is determined by an evaluation of the goodness and/or badness of its consequences. On a *deontological* (nonconsequentialist) approach, rightness or wrongness of an act is determined by considerations other than consequences.

The two main teleological approaches are the **utilitarian approach** and the **self-interested approach.** Each approach maintains that one ought to choose that course of action (among alternatives) that can be expected

1 The concept of moral rightness is at the center of both conduct ethics (doing right) and virtue ethics (being a good person). The meaning of rightness in the former context (conduct ethics) is explored here. Its meaning in a virtue ethics context is discussed in the final section of this chapter titled "Virtue Ethics."

to result in the best consequences for an identified reference group. For a utilitarian, the only acceptable reference group is the class of all people whose well-being stands to be affected. From a self-interested point of view, the reference point in terms of which consequences are evaluated may be oneself as an individual or "one's own" (e.g., one's family, one's business, one's political party, one's nation, one's species).

Utilitarianism. The principle of utility has been conceptualized and applied in a variety of ways. The root idea is that the moral adequacy of an act or rule is determined by its "utility" in maximizing well-being in peoples' lives. A distinction is made between "act utilitarianism" and "rule utilitarianism" An act utilitarian argues that the principle of utility should be used as a gauge to assess the moral justifiability of particular *acts*, in which the well-being of all parties that stand to be affected by that act are taken into consideration. A rule utilitarian argues that the principle of utility is more appropriately applied to *rules* (rather than acts), in which the well-being of society as a whole becomes the focus of the evaluation. Differences arise in connection with how "well-being" is defined. To be applicable in utilitarian deliberation and critique, the definition must specify what it is that should be maximized, i.e., an "objective function" needs to be specified in terms of which the utility of an action or practice is to be measured. There are also differences among utilitarian theorists as to whether total or average utility should be maximized, or whether the objective should be the minimization of disvalue (costs, harms, disadvantages).

Self-interest. Self-interest is, its defenders maintain, a rational and psychologically adequate view to take, and is a view that virtually all individuals and organizations do in fact take. Of course, an individual should do what he or she believes to be in his or her best interests. Likewise, a mother and father should do what is in the best interests of their children, a corporate CEO should do what is in the best interests of shareholders, and elected officials should do what is in the best interests of the constituencies they serve. Whatever the reference point may be (an individual, a family, a business, a nation), it is emphasized that self-interest ought to be pursued in an "enlightened" manner: the interests of other parties should be taken into consideration and common or overlapping interests should be strategically identified in determining how one's own interests can most effectively be served. But when "push comes to shove" and conflicting or competing interests cannot be resolved or harmonized, one should do what one considers to be in one's own best interests in the long term. Self-interest is labeled as "egoistic," but an enlightened egoism goes far beyond narrow selfishness.

There are numerous **deontological frames of reference** for thinking about moral rightness and wrongness, including the following:

Ethical relativism. An ethical relativist associates moral rightness with what people in a given culture or society believe is right as reflected by that society's customs, traditions, mores, and laws. On this approach, one's most general moral obligation is to act in accordance with what is believed to be right in the society of which one is member or a visitor. ('When in Rome, do as the Romans do.")

Rights-based theory. Moral rights theory is discussed at some length subsequently. Distinctions are made between negative and positive rights, and between universal and conventional rights. For theorists that argue certain rights are inalienable and applicable to all relevant parties at all times and places, the basic obligation is to champion rights recognition and compliance."

Religious ethics. Righteous living has been viewed through the lens of religious beliefs for centuries. In Islam, righteous living means following the teachings of the Qur'an and the prophet Mohammed. In Roman Catholicism, prescriptions for living are codified in the Catholic Canon. In the world's religious traditions in general, the basic obligation is to do what divinity reveals to be right through scripture, ordained teachers, and/or divine inspiration.

Kantian ethics. In his *Metaphysics of Morals*, the influential German philosopher, Immanuel Kant, associated the moral rightness of a maxim ("a motive or internal rule of action") with two basic requirements of rationality: consistent universalizability and respect for rationality. In regard to the first requirement, the test of the moral rightness or wrongness of a maxim is whether it can be consistently universalized. In regard to the second requirement, the test is whether it involves treating any rational human being as an end and never as a means only.

Social contract theory. In social contract theories of obligation, moral rightness is associated with institutional arrangements and rules of conduct that a person, thinking rationally, would give one's intellectual assent to and agree to live by; one ought to abide the rules of a "social contract" hypothetically (if not actually) entered into. John Rawls is a prominent social contract philosopher whose view on obligations owed to future generations is discussed in Chapter 10.

Normative proceduralism. Here moral rightness is associated with what people come to agree upon (ideally unanimously) when procedural norms of discussion are followed in sincere attempts to reach agreement—when, for example, the discussion leading to agreement is free of deceit and coercion, inclusive, sustained, open-ended, and factually informed.

Value: Constituent Elements of Well-Being

Another basic moral concept pertains to what constitutes value (goodness, well-being) in the lives of individuals and society at large. For teleological theories of moral rightness—theories that associate rightness and wrongness with the goodness and/or badness of consequences—a method of evaluating consequences is essential.

Various approaches are used to determine what course of action (among alternatives) has the *best consequences*. Commonly used methodologies for determining "best consequences" from both self-interested and utilitarian points of view include the following:

Hedonistic approach. Here the objective is to *maximize happiness* in one's own or other peoples' lives, where happiness is defined in hedonistic terms as the *greatest balance of pleasure over pain.*[2] The hedonistic utilitarian, Jeremy Bentham, proposed a "hedonistic calculus" as a means for determining what course of action, among alternatives, has the best consequences.

Needs-based approach. Here the objective is to *maximize need fulfillment*. When the most basic physiological needs of people are unmet and high levels of homelessness, malnutrition, disease, and premature death are prevalent, priority is given by a needs-based utilitarian to the alleviation of such conditions.

Preference-based approach. Here the objective is to *maximize preference satisfaction*. Preferences are presumed to reflect needs and what makes people happy, and sophisticated methods for identifying and measuring preferences have been developed and are widely applied: surveys, voting, polling, and the analysis of buying behavior. Political and marketing campaigns are based

2 There are, of course, other conceptions of what constitutes happiness in human lives. The Aristotelian conception is discussed subsequently in this chapter in the section titled "Virtue Ethics."

on analyses of people's preferences, as are laws enacted by legislatures and administered by governmental agencies.

Cost-benefit analysis. Here the objective is to gain the *greatest return on investment* or realize the *highest calculated benefit/cost ratio*. Decision making in business corporations and government agencies is primarily based on cost-benefit analysis.

These and other value orientations will be explored throughout this work, and attention will be given to how differing value assumptions and arguments among anthropocentric, vertebracentric, biocentric, and ecocentric theorists play out in debates and policy proposals. A major part of the discussion in environmental and animal ethics is directed toward the anthropocentric thesis that only humans value, have values, and exhibit value-directed behavior, together with the anthropocentric reduction of all value to human value and the further reduction in cost-benefit analysis of all human value to monetary value.

As part of this discussion, a number of value distinctions are made. One is the distinction between intrinsic value and instrumental value. Something is said to have **intrinsic value** (to be intrinsically good) if it is good in and of itself, good for its own sake. In human ethics, pleasure (the experience of pleasure), achievement (in an important endeavor), and excellence (in a worthwhile activity) are singled out as significant intrinsic values. Something is said to have **instrumental value** (to be extrinsically good) if it is good as a means to some other value.

Another value term—**inherent value**—has become an integral part of the discussion whether valuational processes occur independently of humans in the natural order. Other terms used in such contexts are **constitutive value** and **systemic value**. Self-actualization is said to constitute the good of individual organisms; genomic expression is said to constitute the good of both living individuals and species as a whole; mutually advantageous relations (adapted fitness) among populations of many species-kinds are said to constitute the good (stability, integrity) of ecosystems.

Moral Rights: Entitlements to Certain Kinds of Treatment

Moral arguments for social change have been framed using the concept and language of rights, and have fueled important social movements. The pronouncement in the US Declaration of Independence that all men have "certain inalienable rights, among them life, liberty, and the pursuit of happiness" led to the creation of a new nation and the US constitutional system of government. Insistence that all people have a right not to be enslaved led to the abolition of slavery. Other basic human rights claims—the right of women to vote, the right of workers to form unions and engage in lawful union activities, the right of all citizens not to be discriminated against on the basis of gender, race, national origin, religious faith, and age—have led and continue to lead to major institutional changes in American society.

A distinction is made between universal and conventional rights. **Universal rights** have been characterized as "human rights," "inalienable rights," and "natural rights": rights whose legitimacy does not depend on their recognition or protection in law. In 1948, the General Assembly of the newly created United Nations unanimously adopted a Declaration of Human Rights. Such rights have been variously claimed to be God-given (divine law), fitting and proper in the natural order of things (natural law), and/or intuitively self-evident. **Conventional rights**, in contrast, are established by convention and derive their validity from the mores, customs, traditions, and laws of a given society. Ethical relativism acknowledges conventional rights but denies the legitimacy of universal human rights.

Another useful distinction drawn in connection with a discussion of moral rights is that between negative and positive rights. **Negative rights** are entitlement claims not to be harmed in certain ways (e.g., the right not to be enslaved, tortured, discriminated against, slandered, and prevented from freely speaking one's mind). **Positive rights** are entitlement claims to certain benefits (e.g., the right of people living in property to certain kinds of public assistance, the right to medical services in a medical emergency, and the right to a minimum wage).

The United Nations Declaration of Human Rights contains a mix of negative and positive rights. Article 2 ("Everyone has the right to life, liberty, and security of person") and Article 4 ("No one shall be held in slavery or servitude; slavery and the slave trade shall be prohibited in all their forms") are acknowledged to be negative rights. Article 25 ("Everyone has the right to a standard of living for the health and well-being of himself and his family ... and the right to security in the event of unemployment, sickness, disability, widowhood, old age or other lack of livelihood in circumstances beyond his control") is acknowledged to be a positive right. It is widely accepted that negative rights, but not positive rights, impose strict duties not to violate the rights in question and that penalties for noncompliance are justifiable. In this connection, it is argued that Article 25 does not impose a strict duty on governments of underdeveloped nations to implement measures that guarantee standards of living comparable to those of developed nations, or that internationally imposed penalties for lagging standards of living in poor and underdeveloped nations are morally justified. Instead, the Article 25 and other positive rights championed in the UN Declaration are interpreted as ideals to be achieved or standards against which progress can be gauged.

Arguments for the recognition of nonhuman rights are addressed in subsequent chapters. Christopher Stone makes the case for expanded recognition of nonhuman legal rights, noting that "whereas only landowning white adult males enjoyed legal rights two centuries ago in the US, legal standing now extends to non-landowners, women, blacks, Native Americans, and such things as corporations, trusts, cities, and nations." In Stone's view, the time has come to extend legal rights protection to nonhuman entities and for courts to appoint guardian-attorneys to represent their interests in national and international courts of law.

Justice: Fairness in the Treatment of Individuals

The terms "justice" and "fairness" have multiple meanings and pertain to questions concerning how risks, benefits, and costs should be distributed in society (distributive justice), how wrongdoers should be treated (punitive, rehabilitative, or deterrent justice), and how the victims of wrongdoing should be treated (compensatory or restitutive justice).

In regard to distributive justice, there are numerous contexts in which the allocation of benefits and burdens are relevant and oftentimes hotly debated: wealth in an economy, pay and promotions in workplaces, admissions to universities, grades in a classroom, chores in a family, welfare benefits to people living in poverty, and taxes imposed to finance government. Whenever people come together in cooperative social arrangements, things get distributed and some people fare better than others. Is it fair that the wealthiest 1 percent of the population owns about 40 percent of total national net worth? Is it fair that average earnings of the CEOs of major corporations are more than two hundred times more than average earnings of corporate employees? Is it fair that Americans comprise less than 4 percent of the world's population, but generate more than 30 percent of the world's pollution and wastes? Different and oftentimes conflicting theories of distributive justice address such questions.

Moral Standing: Who or What Counts Morally

To assert that a biotic entity has moral standing is to assert

- that the entity in question is morally considerable in its own right; [3]
- that the question of the moral justifiability of the harm in question is a morally relevant question; and/or
- that the entity is the kind of entity to which a moral obligation could reasonably be said to be directly owed, or to which a moral right could reasonably be ascribed.

Although tribal sentiments have led some groups throughout human history to deny moral standing to other humans, most people today readily acknowledge that all humans count morally and that it matters morally how humans treat other humans. But what about other biotic entities? What about nonhuman primates, mammals, or other sentient animals? What about invertebrate animals, insects, or plants? What about collective entities such as species and ecosystems? Do any of these count morally *in their own right*?

Various positions have been staked out on the question of moral standing. ***Anthropocentric*** theorists contend that all humans and only humans count morally. Theorists in the ***vertebracentric*** camp generally contend that all vertebrate animals also count morally, though some of these writers limit nonhuman moral standing to mammals or primates. ***Biocentric*** theorists go further, arguing that all living beings count morally, with some biocentrists (not all) claiming that species have moral standing, not only their living individual members. **Ecocentric** theorists go further still, arguing that not only the living inhabitants of ecosystems count morally but also ecosystems as a whole. Each of these positions will be examined in subsequent chapters.

For the sake of clarity, a distinction needs to be made between **moral standing** and **moral significance**. While a claim of moral standing is a claim that an entity is morally considerable in its own right, it is not a claim that the entity itself is worth more or less than some other morally considerable entity. While the entity must possess something that makes it worthy of the minimal level of moral respect involved in the recognition of moral standing, the concept of moral standing is silent on the question of the relative moral

3 The predicates "has moral standing," "is morally considerable in its own right," and "counts morally" are regarded as synonymous in meaning and are used interchangeably.

worth of various biotic kinds—whether, for example, a human is worth more than an animal, an animal is worth more than a plant, or a wild animal is worth more than a domesticated animal.

A claim that an entity has moral standing is also not a claim that human-inflicted *harm* to the entity in question is of more or less moral *significance* than comparable harm to another morally considerable entity, or that harm to the entity in question should be given more weight in moral deliberation than comparable harm to another morally considerable entity. If, for example, an animal is harmed by a human to secure a human benefit, and both the animal and the human are deemed to have moral standing, the moral standing of the animal would have no bearing on how the harm (to the animal) and the benefit (to the human) should be value-weighted in a deliberation concerning the moral justifiability of the human-inflicted harm involved.

DECISION PRINCIPLES

For the purposes of facilitating quick reference, the following lists moral principles frequently invoked in moral deliberation and critique in human, environmental, and animal ethics. While each of these principles are given tailored formulations depending on context and theoretical orientation, the following formulations reflect common usage.

- **Principle of self-interest**. One ought to choose and act in ways that can reasonably be expected to result in the greatest good for a specific person or group.
- **Principle of utility**. One ought to choose and act in ways that can reasonably be expected to result in the greatest balance of good over bad in the lives of all affected parties or society as a whole, in which like interests among affected parties are given like consideration (equal weight) in moral deliberation.[4]

---

4 The principle of utility was articulated by Jeremy Bentham and John Stuart Mill in the mid-nineteenth century in England. Peter Singer, one of the most influential champions of animal welfare, is a contemporary utilitarian in the tradition of Bentham and Mill. The principle of utility, as stated above, incorporates the principle of equal consideration, which presupposes the principle of equality.

- **Principle of nonmaleficence.** It is wrong to inflict unnecessary (avoidable) harm intentionally and knowingly. Versions of this principle can be found in virtually all ethical traditions, including the Hippocratic oath (dating from the fourth century BC) in which new physicians pledge "before the gods of medicine" to "First do no harm."
- **Principle of equality.** All persons have equal intrinsic worth; no person is inherently superior or inferior in worth or dignity to any other person.
- **Principle of equal consideration.** Like interests ought to be given like consideration in moral deliberation; one person's interest in not being harmed (in a specific way) ought to be given the same weight as any other person's interest in not being harmed (in that way).
- **Principle of precaution.** If significant uncertainty exists concerning the severity of risk associated with an undertaking, action ought to be delayed until it can confidently be judged that the undertaking is acceptably safe.
- **Principle of sustainability.** Activities involving levels (rates) of production and/or resource utilization ought not exceed levels (rates) necessary to enable continuation of those activities over a long planning horizon; production and/or resource utilization ought to be sustainable in the long term.[5]

VIRTUE ETHICS

Virtue Characterized

A virtuous ethicist contends that a morally right action in a given situation is one that a virtuous person, acting in character, would do in that situation. While acknowledging that decision principles are needed to guide action through all of life's challenges and contingencies, the virtue ethicist emphasizes that dispositions to think and act in morally right ways are what is most basic and needed. For the virtue ethicist, the question "What kind of a

5 Versions of the sustainability principle formulated by ecologically oriented resource economists such as John Daly and William Rees are reviewed in Chapter 10 along with a review of sustainability proclamations adopted by the European Union, the World Bank, and various UN organizations.

person would do that?" is no less basic and certainly no less telling than the question "Why would, or why did, you do that?"

Virtues are variously characterized as what makes one a morally good person, as good character traits, as dispositions to act in morally right ways, as the ethical core of personality, or, metaphorically, as "habits of the heart." The virtue ethicist is quick to note that virtues are more than mere habits or dispositions. The "something more" is the value that is attributed to certain tendencies, habits, or dispositions. Virtues are good (valued) character traits; vices are bad character traits. Kindness is a virtue, cruelty a vice; honesty is a virtue, dishonesty a vice; humility is a virtue, arrogance a vice. Habits don't define a person morally, but virtuous dispositions do.

Virtue ethicists emphasize, moreover, that virtuous dispositions do not come ready made but are acquired through experience and are initially cultivated by parents, teachers, and good role models in the formative years of personality and character development. It is emphasized, moreover, that the process of character formation continues throughout life: a process in which virtuous dispositions are broadened and deepened through experience and continuing education and training. Positive enforcement fosters the cultivation of positive virtues and negative enforcement ("learning the hard way") discourages the entrenchment of wrongful tendencies (vices). It is further emphasized that feeling is as much a part of virtue as intellectual comprehension, and that emotional maturity and the ability to control impulses is as much a part of *being* virtuous as knowing right from wrong.

Aristotelian Virtue Ethics

It is instructive to recount how the elements of virtue ethics previously noted have their origins in Aristotelian virtue ethics.

For Aristotle, the ability of humans to reason is the defining essence of humankind and what distinguishes humans from plants and animals. Reason has both a theoretical and a practical component. The end of theoretical reason is knowledge (*gnosis*), gained as one comes to acquire observational and reasoning skills and an understanding of the "first principles" of physics and metaphysics. The end of practical reason is wisdom (*phronesis*). The man who achieves theoretical knowledge and practical wisdom is the man who reasons with excellence (*arête*).

For Aristotle, the highest human good is "living well" (*eudaimonia*): living a life informed and directed by excellence in reasoning. Only *eudaimonia* is worthwhile for its own sake. Wealth, power, and fame may be good as means to other ends, but they do not qualify as ends worthy in themselves.

Eudaimonia is sometimes translated as "happiness," but not happiness in the hedonistic sense of this term (pleasant feeling). Similar to wealth, fame, and power, pleasure is not a worthy human end. Happiness in its most human, satisfying, and enduring form comes from *eudaimonia* (living well). Aristotle introduces the "hedonistic paradox" in this context: a person who is engaged in a single-minded pursuit of pleasure for its own sake paradoxically fails to find it; pleasure cannot be disassociated from the activities that produce it. Unlike elusive, short-term pleasure, happiness in the Aristotelian meaning of this term is the contentment or sense of well-being that "accompanies" living well (*eudaimonia*).

Excellence (*arête*) in practical reasoning involves an ability to discern the middle path (the mean) between extremes in situations in which choices have to be made. Here we have Aristotle's Doctrine of Golden Mean. With respect to certain types of situations, a particular moral virtue is a disposition to discern the middle path (the mean) between excess and deficiency in situations in which a decision must be made. In dangerous situations, for example, *courage* is the mean between the extremes of cowardice (paralyzing fear) and reckless overconfidence (utter lack of fear). In situations in which one is able to share what one possesses with another person in need, *generosity* is a mean between the extremes of stinginess and profligacy. In regard to pleasurable activities, *temperance* is a mean between the extremes of overindulgence and utter lack of participative desire. Consistent with this "middle path" rationale, Aristotle developed a detailed account of numerous virtues.

Consistency in the ability to discern and stay on the "middle path" in choice situations is the hallmark of practical wisdom (*phronesis*), and is an excellence acquired from teaching, for example, and the successes and failures of everyday living. A truly wise person is a person who has acquired the constellation of moral virtues that enables one to do "the right thing, in the right situation, to the right person, at the right time": an ability that is as rare as it is laudable.

Environmental Virtue Ethics

Chapter 4 discusses contributing elements to an attitude of respect toward nature, and concludes with a discussion of four primary environmental virtues: respect, humility, gratitude, and responsibility. As an overview of recent work in environmental virtue ethics and to provide context for this subsequent discussion, the following highlights recent work in the field.

In her book, *Dirty Virtues*,[6] Louke van Wensveen observes that there is a seeming reticence on the part of many ecowriters today to identify with the virtue ethics tradition of the past. She finds this to be somewhat paradoxical given that she is "yet to come across a piece of ecologically sensitive philosophy, theology, or ethics that does not in some way incorporate virtue language." The language of virtue "turns up in the writings of social ecologists as well as deep ecologists, bioregionalists as well as animal rights activists, creation theologians as well as environmental philosophers, mainstream theologians as well as radical ecofeminists." What, then, she asks, is the explanation for this "virtue blindness"? One reason is that ethics in general over the past two centuries has focused on rights, values, duties, principles, and consequentialist arguments, with relatively little attention given to virtue ethics. A second reason is that renewed interest in virtue ethics occurred in the middle part of the twentieth century, about the same time that environmental ethics began to take shape as a new field of inquiry, with the two developing more or less independently. A third reason is that "the terms 'virtue' and 'vice' carry the stigma of sounding old-fashioned, preachy, and self-righteous."

Despite the lack of outright identification with past traditions of virtue ethics, Van Wensveen goes on to explain why one nevertheless finds the use of virtue terminology to be widespread. It is, she says, because many ecologically minded people are very worried about the environmental and ecological devastation that they see all around them, and because virtue ethics provides a vocabulary for expressing this concern and a commitment to address it. Furthermore, virtue discourse allows and encourages the integration of emotions, thoughts, and actions; it fits with the ideal of personal wholeness that many ecologically minded people espouse; it depends on narratives, vision, and the power of examples; it fits with the preference of many ecologically

6 Louke van Wensveen, *Dirty Virtues: The Emergence of Ecological Virtue Ethics* (Amherst, N.Y: Prometheus Books, 2000).

minded people for change through conviction rather than coercion; and it provides rich expression to many different facets of human experience.

Van Wensveen then makes a number of observations about the characteristics of environmental discourse and the role it plays in the field of environmental ethics more generally. Her first observation is that virtue ethics discourse is *integral.* Terminology connoting virtue (e.g., respect, adaptability, benevolence, care, creativity, cooperation, nurture, frugality, humility, gratitude, hope, moderation, discipline, restraint, openness, perseverance, wisdom) and vice (e.g., arrogance, carelessness, competitiveness, greed, contempt, cruelty, domination, indifference, insensitivity, pride, haste, wastefulness) are used to articulate integrated perspectives. Her second observation is that virtue discourse is *diverse*, revealing differences in the worldviews, backgrounds, and characters of authors in the field. This diversity shows up in the configuration of virtues and vices used and the emphasis given to each. She notes in this regard that there is sometimes outright disagreement; one writer's version of stewardship may be found to be presumptuous or to reflect a lack of sensitivity in the mind of another writer, or what may be a particularly worrisome vice to one author may seem a small or misplaced concern to another. Her third observation is that virtue discourse is *dialectical* and creates tensions. While some tensions reflect the infancy of the discipline in need of being straightened out, others, she says, reflect a dialectical richness that ensures flexibility and prevents harmful extremes. Her fourth observation is that ecological discourse is *dynamic.* The language is emerging and changes with changing circumstances and discoveries in science. Her final observation is that virtue discourse is, for the most part, *discourse without a social ethic.* While virtue discourse is intimately tied to visions of human and social ideals and conveys commitment to change, writers in this tradition rarely present developed social change theories. In her view, the discourse "needs to become more integrated with those sections of environmental literature where we do find the development of a social ethic."

In the introduction to *Environmental Virtue Ethics*,[7] an anthology of papers coedited by Ronald Sandler and Philip Cafaro, Sandler comments that the environmental virtue literature has revolved around two basic

---

7 Ronald Sandler and Philip Cafaro, *Environmental Virtue Ethics*, Ronald Sandler and Philip Cafaro, eds. (New York: Rowman & Littlefield, 2005). The authors of this anthology include ten commissioned chapters and four previously published works in the field.

unity of dispositions to act and feel for certain sorts of reasons, and asking whether this complex unity would be the sort of thing that could be inculcated in children as part of their moral education. Notwithstanding the complexities, Hurtshouse envisages two promising candidate new environmental virtues: *wonder* and *respect for nature*.

In regard to *wonder*, she comments that "the putative virtue of being disposed to feel the emotion of wonder the right way, towards the right objects, for the right reasons, to the right degree, and so on is, I think, explicitly concerned with our relations to nature." In regard to *respect for nature*, Hurtshouse is highly critical of Paul Taylor's treatment of this concept, which associates respect for nature with a set of beliefs that he (Taylor) calls the "biocentric outlook."[9] Her critique begins by emphasizing that "adopting" or "taking up" a set of beliefs is a far cry from "being rightly oriented to nature, through and through, in action, emotion, perception, sensibility, and understanding." In her view, "That is the sort of change that cannot come about just as through, say, reading a philosophical book and deciding to change; it cannot simply be 'adopted' or 'taken up.'" Hurtshouse also objects to Taylor's sharp distinction between animate and inanimate nature. "But," she continues, "if we think of being 'rightly oriented to nature,' not as an attitude founded on an adult's rational recognition of a one-sentence premise but as a character trait arising from a childhood training that gives us particular reasons for action in particular contexts, and shapes our emotional response of wonder, the hard and fast line [Taylor] draws between the animate and the inanimate becomes insignificant." As a parting shot, Hurtshouse comments that the normative principles developed in Taylor's biocentric ethics offers only "the obvious prohibitions that even the palest green environmentalist is already living."

9 Taylor's biocentric is discussed in some detail in Chapter 7.

2 ENVIRONMENTAL AND ECOLOGICAL DECLINE

To provide context for the subsequent discussion of anthropocentric, vertebracentric, biocentric, and ecocentric perspectives in environmental and animal ethics, and to highlight the seriousness of the issues addressed, this chapter presents an overview of the many ways in which we are living in a period of environmental and ecological decline.

The signs of decline are evident on many fronts: greenhouse gases are concentrating, global temperature is rising, glaciers and polar ice are melting, sea levels are rising, oceans are acidifying, extreme weather events are intensifying and occurring more often (hurricanes, tornadoes, wildfires, extreme heat and cold, protracted droughts), coral reefs and tropical rainforests are vanishing, boreal forests are fragmenting and diminishing, groundwater availability is decreasing, deserts are advancing, and pollutants and wastes are multiplying. On top of this, our bodies are loaded with toxins (one study determined that the umbilical cords of Chicago-area newborns contained one hundred ninety toxins), and trillions of animals are held in captivity in factory farms, laboratories, and zoos, and treated in ways that involve untold suffering.

This chapter begins with a discussion of several macro-level indices of unsustainability, followed by a discussion of the underlying causes ("drivers") of the paths of unsustainability we are on. The focus then turns to a more in-depth discussion of three of the most serious threats facing peoples around the world: climate change, biodiversity loss, and pollution/waste generation.

INDICES OF UNSUSTAINABILITY

One high-level measure of unsustainability is the proportion of total planetary **net primary production** (NPP) appropriated by humans. NPP is the energy generated by photosynthesizing plants and cyanobacteria in excess of that required for the functioning of those organisms, with the surplus available for other life.[1] Prior to the ramp-up of the Industrial Revolution, it is estimated that humans appropriated about 30 percent of total planetary NPP. Today, humans co-opt about 40 percent of total NPP, a proportion that could increase to 50 percent or more if the human population reaches twelve to fifteen billion. Growing human appropriation of Earth's NNP means less energy for other life and ultimately the stock of NPP producers: an inherently unsustainable process.

Another widely used indicator of unsustainability is the size of the **ecological footprint** of a "center" of resource consumption (e.g., a particular person, a city, a region, a nation, or humanity as a whole). William Rees, an ecologist who coined the term, characterized an ecological footprint of a region as a measure of the extent to which regionally available resources are insufficient to sustain that region's current level of production and consumption, i.e. a measure of the extent to which imported resources from elsewhere are required. Rees estimated that the ecological footprint of the Vancouver, BC, region is twelve times greater than the territory it occupies, "Having overused or overdrawn local natural capital stocks, the people of this region are heavily dependent on *temporary* surpluses elsewhere." In metropolitan areas with far less locally available resources (e.g., Los Angeles), the ecological

1 The importance of photosynthesizing plants needs no elaboration here, but the role of cyanobacteria in the biosphere is less well known. Cyanobacteria are thought to contribute from 20 to 30 percent of the planet's photosynthetic capacity, and ancient cyanobacteria are believed to have led to the evolution of plastids in plants. Using solar energy, photoautotrophic cyanobacteria split water molecules into oxygen, protons, and electrons, and use the high-energy electrons for their own functioning, with any surplus being available for other life. As put by biologists Stewart and Falconerm: "Cyanobacteria are arguably the most successful group of microorganisms on earth. They are the most genetically diverse; they occupy a broad range of habitats across all latitudes and are widespread in freshwater, marine, and terrestrial ecosystems, even in the most extreme niches such as hot springs, salt works, and hypersaline estuaries. Oxygen-producing cyanobacteria created the conditions in the planet's early atmosphere that directed the evolution of aerobic metabolism and eukaryotic photosynthesis. Cyanobacteria fulfill vital ecological functions in the world's oceans, being important contributors to global carbon and nitrogen budgets."

footprint is much larger. Rees continues: "Our technological prowess feeds the common belief that humans and their economies have become increasingly independent of nature; in actuality, however, modern high-income economies are becoming increasingly indebted to nature and are running massive unsustainable ecological deficits."[2]

The Global Footprint Network maintains a widely used methodology for estimating the ecological footprints of individuals, cities, and nations. In this approach, the ecological footprint of a region is developed by summing the hectares of cropland, forestland, grazing land, fishing ground, and built-up land to produce the amount of food, fiber, and timber regionally consumed, together with the amount of carbon dioxide emitted from these activities. This footprint is then compared with a biocapacity measure derived by summarizing the various land and aquatic resources (again measured in hectares) that are available globally to meet demand. Using this model, ecological footprints range from a 0.4 hectares/person/year for a person living in Bangladesh, to a footprint of 8.04 hectares/person/year for the United States and 10.5 hectares/person/year for the United Arab Emirates.

On a global scale, the World Wide Fund for Nature reports in its 2018 *Living Planet Report* that the bio-regenerative capacity of the planet has already been substantially exceeded: that 1.6 Earths are required to support current worldwide demand on natural resources. If everyone had the ecological footprint of an average American, 3.9 Earths would be required; for an average person currently living in Qatar, the requirement would be 4.8 Earths.

Most people are oblivious to the unsustainability of current lifestyles and economies for two main reasons: because people in many poor and underdeveloped nations consume resources at rates far lower than people living in wealthy developed nations, and because people have hitherto been able to cope with shortages through trade, exploitation of finite resources built up in the past (e.g., fossil fuels), and borrowing from the future (e.g., overexploitation of forests and fisheries). But the time has arrived when such hedges against ecological deficits cannot be depended upon. As Herman Daly explains, growth becomes uneconomic and eventually impossible when consumption of renewable resources exceeds renewal rates, when rates of

2 William E. Rees, "Consuming the Earth: The Biophysics of Sustainability," in *Earth Ethics*, James P. Sterba, ed. (Upper Saddle River, NJ: Prentice Hall, 2000), pp. 382–386.

consumption of nonrenewable resources exceed rates at which technology enables the development of resource substitutes, and when rates of waste generation exceed the capacity of ecosystems to absorb and recycle those wastes.

The concepts of sustainability and ecological footprints will be explored at greater length in Chapter 10.

UNDERLYING CAUSES

In his book, *Red Sky in the Morning*,[3] Gus Speth singles out population growth, economic growth, market failure, and political failure as the underlying causes of environmental and ecological decline.

Population Growth

Population growth increased fourfold during the twentieth century. In calendar year 0, it is estimated that there were about five hundred million people in the world. By 1750, the size of the human population is estimated to have grown to about one billion, to three billion by 1960, and to six billion by 2000. It took 1,750 years to get the first doubling of the human population, but only forty years, between 1960 and 2000, to get another doubling.

Today the human population worldwide stands at about 6.72 billion. While population numbers and rates vary from year to year and decade to decade, the overall growth rate has tapered in recent decades, with much of the increase occurring in less-developed nations. The United Nation's Intergovernmental Panel on Climate Change projects that the human population will continue to expand, reaching about 8.5 billion in 2030, 9.0 billion by 2050, and 11.2 billion by 2100.

Population is the "elephant in the room" that policy makers are hesitant to address in many policy forums because of opposing and emotionally charged stances on controversial issues such as birth control, reproductive rights, and women's rights to choose. Without wading into these waters, suffice it to say here that the already large and growing number of people in

3 Gus Speth, *Red Sky at Morning* (Yale University Press, 2006).

the world are putting immense pressure on Earth's limited and dwindling resource base—pressure that will only increase as people living in under-developed nations (representing more than half of the world's population) aspire to standards of living currently enjoyed by people living in developed nations.

Economic Growth

While the world's human population increased fourfold during the twentieth century, economic growth (total world GDP) increased twentyfold. From the dawn of civilization to 1950, Speth states, the world economy grew to seven trillion dollars and now grows by this amount every five to ten years.

As noted earlier, continued growth of such magnitude is seen by Rees, Daly, and other economists and ecologists as unsustainable. Economic theories are advanced that challenge assumptions in classical resource economics, notably the assumptions that the resources needed for continuing economic growth exist in abundance and are replenished and renewed by processes in the natural world, that the natural environment is able to safely absorb waste by-products and pollutants without diminishing regenerative capacities, and that technology can be relied upon to develop resource substitutes when critical resources are in short supply.

Calls come from many quarters for lower levels of growth and for abandoning altogether the notion that growth is an inherent and desirable feature of a viable economy. In political and activists circles, many parties embrace the "Global Greens Charter" decrying the "dogma of economic growth at any cost": a dogma leading to "excessive and wasteful use of natural resources ... and extreme deterioration in the environment and massive extinction of species."

As emphasized by the Intergovernmental Panel on Climate Change, stabilization of heat-trapping carbon dioxide in the atmosphere—a return to the concentration of atmospheric carbon that prevailed for some ten thousand years prior to the ramp-up of the Industrial Revolution—will be possible only if population and economic growth are stabilized, together with the near-term transitioning from fossil fuel to renewable energy sources in the world's economies.

Market Failure

A third underlying cause of environmental and ecological decline singled out by Speth is the failure of markets and businesses to account for all of the costs associated with the production of economic goods, including the costs associated with the appropriation, depletion, and/or degradation of natural capital, i.e., the myriad ecological services that make human and all other life possible, including net primary production by photosynthesizing organisms, soil making, nutrient recycling, pollination, aquifer recharge, and air and water cleansing, Such services are regarded in classical economics as "free goods." They enable businesses to make profits and create wealth, but they are not treated as costs; businesses do not pay for them; they do not show up as costs in accounting systems; and they are not reflected in the prices charged for goods and services.

Businesses pay for the internal costs of doing business—supplies, labor, machinery, interest on loans, taxes, and the like—but they do not pay for costs incurred by consumers injured or sickened by the products they produce (unless court-ordered liability payments are made), nor do they pay for the costs associated with by-product pollution and waste generation (unless fines, disposal costs, or special taxes are involved). Victims and/or insurance companies typically pay for product-caused injuries and illness, and taxpayers typically pay for environmental costs in the form of taxes to fund government regulatory agencies.

Businesses do not pay for the ecological services on which they depend, nor do they typically pay for ways in which they deplete or degrade ecosystem functioning and diminish the capacity of ecosystems to regenerate and continue providing essential ecological services. Government policies aimed at making businesses pay are addressed in Chapter 9.

Political Failure

Political failure, the fourth major driver singled out by Speth, refers to the failure of governments to correct market failure, encourage environmentally technology, and cooperate with one another and industry in problem-solving efforts.

While encouraging progress is being made on these several fronts—including, notably, concerted efforts under United Nations conventions on both climate change and biodiversity and the embrace and application of zero-waste practices in municipalities, commerce, industry, and institutions—much remains to be done. Also encouraging are steps taken by government to remedy "market failure" and encourage approaches designed to harness market forces in support of environmental objectives.

CLIMATE CHANGE

While climate has changed dramatically over Earth's long history—there have reportedly been seven cycles of glacial advance and retreat in just the last six hundred fifty thousand years as a result of fluctuations in solar energy output—there is broad scientific consensus that global warming since the beginning of the Industrial Revolution is due primarily to increasing concentrations of greenhouse gases (carbon dioxide, methane, nitrous oxides, and chlorofluorocarbons)[4] in the atmosphere, and that human activities are primarily responsible for the buildup of these gases.

2013 Report of the Intergovernmental Panel on Climate Change (IPCC)

Reflecting growing levels of concern in society over the potential consequences of global warming, the Intergovernmental Panel on Climate Change (IPCC), [5] working under the auspices of United Nations Conference on Climate Change,[6] issued its fourth report. The headlines of the IPCC report were the following:

- that Earth's average global surface temperature is rising, having increased 0.85 degrees centigrade between 1800 and 2012, with 84 percent of this warming occurring since 1950 (with the twenty warmest years on record having occurred since 1980, ten of which during the previous twelve years);

4 While carbon dioxide is the most prevalent greenhouse gas (78 percent of the total), methane, nitrous oxides, and chlorofluorocarbons are thought to be 25, 300, and 1,000 times more heat-trapping than carbon dioxide.

5 The Intergovernmental Panel on Climate Change is comprised of scientists in numerous disciplines and from numerous research organizations around the world.

6 The Conference on Climate Change was created at the United Nations' Earth Summit, held in Rio de Janeiro in 1998.

- that the warming is primarily anthropogenic, caused largely by the burning of fossil fuels (the primary source of carbon dioxide emissions), ruminant digestion in dairy and beef cattle (a primary source of methane emissions), the manufacture and use of fertilizers in agriculture (a primary source of atmospheric oxides of nitrogen), and widespread use of aerosols and refrigerants (the primary source of chlorofluorocarbons);[7]
- that without concerted efforts worldwide to reduce greenhouse gas emissions, planetary warming will continue, very likely reaching 3.7 degrees centigrade by the end of this century; and
- that warming above 2.0 degrees centigrade would very likely have far-reaching and dangerous consequences, including accelerated melting of mountain glaciers and polar sea ice, rising sea levels (inundating low-lying islands and portions of coastal cities, with trillions spent on mitigation and millions of people becoming climate refugees), increasing acidification and warming of the world's oceans (leading to the decline of coral reefs, seaweed, kelp, shellfish, and net primary production by marine microorganisms), and more frequent and more intense weather events (hurricanes, tornadoes, periods of extreme heat and cold, and prolonged droughts).

The 2015 Paris Agreement

Since 1992, parties to the Convention on Climate Change have met in annual meetings to "stabilize greenhouse gas concentrations in the atmosphere ... and to prevent dangerous anthropogenic interference with the climate system." At the 1997 conference in Kyoto, an initiative called the "Kyoto Protocol" was advanced. The protocol called for emissions reduction commitments to be made in two rounds by individual nations based on the "principle of common but differentiated responsibilities"—the principle that developed countries

7 Two widely accepted scientific conclusions further support the thesis that global warming is human-caused: (1) the concentration of atmospheric carbon increased from 280 to 410 parts per million between 750 and 2010, while the concentration of carbon dioxide is believed to have remained relatively stable for some ten thousand years prior to 1750; and (2) while short-term fluctuations in solar energy occur (the source of all planetary energy), the average annual amount of solar energy reaching Earth has remained relatively stable for the past several hundred years.

should bear the brunt of responsibility for reducing greenhouse gas emissions because they are primarily responsible for the global warming crisis. As it became increasingly evident that the Kyoto Protocol was faltering, a renewed initiative was agreed to at a two-week meeting held in Paris in December 2015. The agreement that emerged (the "Paris Agreement") is generally regarded as a separate instrument, not an amendment to the Kyoto Protocol.

At the Paris meeting, one hundred eighty nations and the European Union agreed to set targets ("nationally determined contributions") for reducing carbon emissions, with the goal of ensuring that post-industrial global warming does not exceed 2.0 degrees centigrade by 2030 and of lowering this ceiling to 1.5 degrees centigrade by 2050. The nationally determined contributions are voluntary and are stated as percentage carbon reductions by 2025, using 2010 as the baseline year. It was also agreed that a global "stocktake" will be made every five years to gauge progress in meeting the Paris goals and to make needed course adjustments. It was also affirmed that wealthy nations, which have emitted the largest amounts of carbon dioxide historically, should contribute to a Green Energy Fund to enable poorer nations to make the transition from fossil to renewable energy and to cope with the effects of climate change.[8]

2018 Report of the Intergovernmental Panel on Climate Change

With growing alarm over the dangers of climate change and studies claiming that pledged carbon reductions under the Paris Agreement are insufficient to meet the 2030 goal of limiting post-industrial global warming to 2.0 degrees centigrade, the IPCC issued an interim stocktake in December 2018. The following highlights conclusions from this report:

- In view of the likelihood that global warming will very likely exceed the 2030 goal, commitments must be stepped up with a focus on ensuring

8 In discussions leading up to the Kyoto conference, various parties contended that national carbon reduction targets should be based on the "principle of common but differentiated responsibilities" noted above: the principle that the level of national emissions reduction targets should be commensurate with the level historical national emissions. This contention proved to be divisive and highly contentious, and was one reason (along with the complexities of attempting to implement a trade-and-cap system for limiting emissions) that led to the collapse of the Kyoto Protocol. The principle was reaffirmed in Paris, however, in connection with discussions over what constitutes equitable contributions to the Green Energy Fund.

that post-industrial warming does not exceed 1.5 degrees centigrade by 2050.

- The risks associated with 2.0-degree warming are substantially greater than those associated with 1.5-degree warming.
- Limiting warming to 1.5 degrees by 2050 will require carbon neutrality (net-zero carbon) by that year by balancing levels of emissions with equivalent levels of carbon sequestration.[9]
- Plans to mitigate expected effects are essential, but not at the expense of relaxing efforts to reduce emissions, thereby diminishing the ability to achieve carbon neutrality by 2050.
- While reductions in carbon emissions are essential, measures must also be taken to reduce emissions of methane, nitrous oxide, and chlorofluorocarbons.
- Emphasis needs to be given to national, regional, and state climate assessments in recognition of differences in local climates and associated local effects.[10]
- Scientific, technological, and financial capacity exists to return to a safe balance between solar radiation entering and exiting the atmosphere.
- Limiting warming to 1.5 degrees centigrade will require everyone—countries, cities, the private sector, individuals—to take immediate and sustained action.

While political will is a critical missing ingredient, growing recognition that renewable energy development will bring large and widespread economic benefits is overcoming the resistance of sceptics. According to *The New Climate Economy*, the report of a study commissioned by the government of seven countries, renewable energy is able to deliver twenty-five trillion dollars in economic benefits, generate more than sixty-five million jobs, and avoid more than 70,00 premature pollution-related deaths. Growing awareness of the potentially catastrophic economic consequences

---

9 Sequestration involves capturing carbon at points of emission and from the atmosphere and storing captured carbon in wells, caverns, and other underground geological formations. It also involves preserving and enhancing the absorption of carbon in soils and vegetation (e.g., extensive planting of trees and tree farms). Proposals are made to enhance ocean absorption of carbon (e.g., by introducing large quantities of iron), but most regard this as a bad idea given the ecological problems associated with increasing ocean acidity.

10 The US National Climate Assessment was released in December 2018. See https://science2017.globalchange.gov/chapter/executive-summary/.

of not reducing dependence on fossil fuels is no less motivating. One recent study concluded that world GDP could be 20–30 percent lower by 2100 if post-industrial warming is not held below 2.0 degrees centigrade: a depression that would be twice as deep as the twentieth century's Great Depression.

LOSS OF BIODIVERSITY: THE HOLOCENE EXTINCTION

We are living in a period of mass extinctions, often referred to as the "Holocene extinction." The following characterizes the extent and seriousness of rapidly declining biodiversity, underlying causes, and debates over conservation priorities.

Scope and Extent

Precise estimates of biodiversity loss are problematic for a number of reasons: because the total number of species is not known with any reasonable degree of confidence,[11] because in many cases the diversity most threatened is also the least explored, and because of the challenges of separating human-caused (anthropogenic) extinction rates from naturally occurring (background) extinction rates. Nevertheless, the scientific consensus is that current rates of extinction match or exceed rates of loss associated with Earth's previous mass extinctions: the Ordovician (440 mya), Devonian (370 mya), Permian-Triassic (245 mya), Triassic-Jurassic (200 mya), and Cretaceous-Paleogene (66 mya). The current period of mass extinctions, the Holocene, has its origins in mid-nineteenth century industrial society and is characterized by steadily growing extinction rates. A notable difference between Holocene extinctions and those of the ancient geological past is that the Holocene involves both plants and animals.

In its 2014 *Living Planet Report*, the World Wildlife Fund (WWF) reported that "the number of mammals, birds, reptiles, amphibians and

11 In *Biodiversity*, E. O. Wilson estimates the total number of currently existing species to be in the neighborhood of twelve million. Some recent studies focusing on insects and microbial organisms suggest significantly higher numbers.

fish across the globe is, on average, about half the size it was 40 years ago." In its updated 2016 report, WWF estimates that 58 percent of vertebrate species, 39 percent of terrestrial wildlife, 39 percent of marine wildlife, and 76 percent of freshwater wildlife have gone extinct since 1970. Reflecting on what is now a consensus view among population and conservation biologists, more than half of species existing in 2000 could become extinct by 2050.

The International Union for the Conservation of Nature (IUCN), a global network of scientists and research stations across the world, publishes annual updates on the status of species using the following categories: extinct, extinct in the wild, critically endangered, endangered, vulnerable, and least concern. In IUCN's 2016 report, 23 percent of vertebrates and 70 percent of plants are red-listed, i.e., are species classified as critically endangered or endangered. Among vertebrates, 23 percent of mammals, 12 percent of all birds, and 32 percent of amphibians are classified as critically endangered or endangered.

In addition to its annually published red lists of endangered species, IUCN also identifies what it calls "biodiversity hotpots," i.e., regions with rapid habitat loss and high concentrations of endemic species (species not found naturally anywhere else). While hotspots are found throughout the world, the majority are tropical forests. Critical hotspots include Brazil's Amazon rainforest, Columbia's rainforest, Madagascar's dry deciduous forests, and Indonesia's seventeen thousand islands.

In regard to invertebrate herpetofauna (insects), researchers at the Australian universities of Sydney and Adelaide and the China Academy of Agricultural Sciences report that "more than 40% of insect species could become extinct within the next few decades" and that "insect biomass is declining at a staggering rate of 2.5% annually." Herpetofauna make up 70 percent of all animal species, and have been at "the structural and functional base of many of the world's ecosystems since their rise about 400 million years ago."

The status of coral reefs (all harboring a rich diversity of life) is no less alarming. At least 27 percent of the coral reefs existing at the turn of the century have collapsed, and virtually all of them will very likely be gone by 2100. Microorganisms, too, are thought to be in rapid decline. Along with insects and fungi, microorganisms play central ecosystem roles as net producers of usable energy in the biosphere and as decomposers and recyclers of dead and decaying material.

Underlying Causes

The causes of species extinctions are numerous and interrelated. The acronym HIPPO—standing for **H**abitat destruction, **I**nvasive species, **P**ollution, human over-**P**opulation, and resource **O**ver-harvesting—identifies the principal drivers in many or most cases.

While the causes of biodiversity decline vary from ecosystem to ecosystem and species to species, habitat loss is, in general, the primary cause. The use of fertilizers, herbicides, and pesticides contribute to significant losses among birds and insects, as do pathogens in many plants and animals.

Altered ecological relations also often play an important role. For example, the rapid decline of insects has cascading ecological effects, including diminishing food sources for other species (60 percent of birds rely on insects as a food source), reduced pollination (80 percent of wild plants use insects for pollination), and substantially altered nutrient recycling in soils. Principal causes of insect decline are habitat loss resulting from the appropriation of forests and wildlands for urban and agricultural expansion, the intensive use of pesticides and fertilizers in agriculture, invasive species, and climate change.

Resource overharvesting occurs when a resource is consumed at an unsustainable rate—witness excessive logging, overhunting, poor soil conservation, illegal wildlife trade, and overfishing. These and other forms of resource exploitation are ultimately driven by growth in human populations and an economic system driven by profit and predicated on the imperative of continuing growth.

Institutional Responses

The United Nations' Convention on Biological Diversity was created at the Earth Summit, held in Rio in 1992, with the charter of "protecting the world's biological diversity, promoting the sustainable use of ecosystem resources, and encouraging an equitable sharing of the benefits arising from genetic resources." Parties to the convention (one hundred ninety-two nations currently) have pledged to develop and implement biodiversity action plans, and to meet every two years to assess progress in meeting biodiversity goals.

Numerous nongovernmental conservation organizations have been established to conduct needed research, build public awareness, and take action, notably the International Union for the Conservation of Nature

(IUCN), the Worldwide Fund for Nature (WWF), The Nature Conservancy, Conservation International, the Worldwide Fund for Nature, the World Wildlife Fund, Ocean Conservancy, and BirdLife International, among numerous others.

Conservation biology has emerged as a major interdisciplinary area of scientific inquiry and is aptly described as a discipline with a mission. The Society for Conservation Biology, a global network of conservation professionals, has more than four thousand participating members worldwide, with the mission of "advancing the science and practice of conserving Earth's biological diversity."

IUCN estimates that 45 percent of Earth's surface are still "wildlands" (defined as lands that retain at least 70 percent of their original vegetation and inhabit less than five humans per kilometer), of which 14.9 percent are currently legally protected. Some 7.3 percent of the world's oceans are also legally protected. Goals set at the 2011 Convention of the Parties (under the Convention on Biological Diversity) call for 17 percent of terrestrial and inland water to be protected by 2020 and 10 percent of coastal and marine waters.

Calls to protect half of Earth for biological diversity date back to the 1970s, but the concept has gained momentum in recent years, largely due to the 2009 founding of the "Nature Needs Half" movement and the 2016 publication of E. O. Wilson's book, *Half Earth*. In December 2018, fourteen nongovernmental conservation organizations issued an online declaration calling for 30 percent of the planet to be "managed by nature" by 2030 and 50 percent by 2050.

Recognition is growing that lands controlled by indigenous people are often much better managed from a biodiversity perspective than lands controlled by settlers. Although indigenous people make up less than 5 percent of the global population, they own or manage about 25 percent of the planet's land. Accordingly, voices behind "30% by 2030" emphasize that indigenous lands should be formally recognized as part of the 30 percent. IUCN has created a classification of protected area, ranging from "strict nature reserves" with limited access for people (Type Ia) to protected areas with sustainable use of natural resources (Type IV). The latter aptly describes lands controlled by indigenous peoples. While discussion continues on land classifications and the meaning of "legally protected," it is most likely that the 30 percent (by 2030) and 50 percent (by 2050) targets will include not only tropical rainforests and set-aside parks and reserves familiar to Americans, but traditional hunting and gathering grounds, natural areas on military bases, areas preserved for scientific research, sacred sites, pastures and native grasslands, and city parks.

Debate over Conservation Priorities

A debated issue in conservation biology concerns the relative significance of the various dimensions of biodiversity loss: loss of biomass (populations) within species, loss of species within ecosystems, loss of ecosystems within regions and the biosphere as a whole, and loss of genetic diversity within populations, species, and ecosystems. The debate centers on the question of priorities and how professional and financial resources can most cost-effectively be allocated.

Early emphasis has been given to the significance of the loss of species, particularly to megafauna to which humans have emotional attachments and/or are teetering on the brink of extinction and/or play key niche roles in ecosystems ("keystone" and "indicator" species). The emphasis on species-centered biodiversity loss stems in large part from IUCN's classification of risk using the designations "extinct," "extinct in the wild," "critically endangered," "endangered," "vulnerable," and "least concern." IUCN's annual "red-listing" of species judged to be at greatest extinction risk, together with its identification of "biodiversity hot spots," further encourages the species-centered approach.

Meanwhile, other voices contend that primary attention should be given to at-risk ecosystems rather than at-risk species: to "cold spots" containing large species-populations (biomass) rather than "hot spots" containing significant numbers of endemic species. Various reasons are given to support this shift in emphasis: (1) Emphasizing biodiversity hot spots tends to downplay social and ecological relationships and overlook areas where biomass, not species diversity, is the most critical factor. It has been noted in this regard that 36 percent of Earth's surface encompasses nearly 39 percent of all vertebrates and that large swaths of biodiversity-rich areas lack significant numbers of endemic species, thus disqualifying them for hot spot attention; (2) Population-level diversity in cold spots are thought to be disappearing ten times faster than diversity at the species level; (3) Geographical areas such as the steppes, taiga, the Serengeti, and the Arctic exhibit population abundance but not an abundance of species; and (4) An ecosystem focus is better suited to assess the critical role that biological diversity plays in the provision of ecological services essential to all life, human and nonhuman. In addition to its annual publication of red lists of endangered species and biodiversity hot spots, IUCN also provides regularly updated lists of organisms by population (biomass).

POLLUTION AND WASTES

Pollution

In regard to growing levels of pollution of the air we breathe, the water we drink, and the sinks to which pollutants are discharged, Speth gave this characterization:

- **From modest to huge quantities:** Pollutants of all kinds have increased dramatically over the past century. Levels of atmospheric carbon dioxide, nitrogen oxide, and sulfur dioxide, for example, increased fivefold, sevenfold, and eightfold, respectively, over the past one hundred years.
- **From gross insults to microtoxicity:** Nineteenth and early twentieth century alarm over city smoke and sewers broadened in the '50s, '60s, and '70s to include the adverse health and economic effects of specific pollutants (e.g., nitrogen oxide in connection with smog, sulfur dioxide in connection with acid rain, phosphorous in connection with eutrophication in bodies of water, and pesticides in connection with the decline of birds and pollinating insects). More recently, the concern has intensified on the dangerous effects of toxins on reproductive, neurological, and immune systems.
- **From First World to Third World:** Concern about air and water pollution in urbanized and highly industrialized nations are painfully evident throughout the world but are often more severe in developing and undeveloped nations, in large measure because major factors contributing to environmental deterioration—poverty, disease, lack of education, and lack of opportunity for women—are found disproportionately in the Third World.
- **From local effects to global effects:** Environmentally and ecologically harmful activities in one area of the world increasingly affect people

and ecosystems in other parts of the world, and cumulative totals and concentrations have global effects, affecting humans all around the world more or less simultaneously.

Solid Wastes

The enormous volume of solid wastes generated around the world annually (disproportionately in wealthy, high-consumption nations), together with carelessness in how these wastes are discarded on the land and discharged into storm drains, streams, lakes, and marine environments, is causing serious environmental, ecological, and health problems.

As reported by the United Nations Environmental Program (UNEP) in its 2009 *Yearbook*, nearly 2.5 billion metric tons of solid waste are generated annually from households, municipalities, commercial establishments, industrial operations, and agriculture. If all 2.5 billion metric tons of solid waste were put on trucks, they would circle the globe twenty-five times. Some of the waste is toxic and classified as hazardous, including chemicals and petroleum products used in manufacturing and agriculture, discarded biomedical material and unused prescriptions, radioactive substances, and explosives. Much of the waste ends up in landfills.

Plastics constitute about 10 percent of solid waste disposal in the United States. Plastic manufacturing and plastic injection molding is the third-largest manufacturing industry, and many types of plastic are used for many purposes ranging from disposable water and soda bottles, to grocery bags and cling wrap, to milk and soap bottles, to vinyl siding and plumbing and electrical conduit.[12] Some five hundred billion plastic bags are used around the world, and most household plastic products are used only once before being discarded—enough to circle the earth four times. Because it takes five hundred to one thousand years for plastic to degrade, virtually every piece of plastic ever made still exists in some shape or form (except for the relatively small portion incinerated).

12 Some of the most commonly used plastic types are *polyethylene terephthalate* used as a fiber in disposable water and soda bottles, outdoor clothing, boats, rope, etc.; *low-density polyethylene* used for plastic grocery bags, plastic cling wrap, sandwich bags, squeeze bottles, etc.; *high-density polyethylene* used for milk, shampoo, soap, determent, and motor oil bottles; and *polyvinyl chloride* used for vinyl siding, plumbing and electrical pipe, outdoor furniture, flooring, etc.

Plastics reaching the world's oceans are particularly problematic for marine wildlife. According to the Ocean Conservancy, plastic constitutes approximately 90 percent of all trash floating on the ocean's surface. These plastics can drift for years before eventually concentrating in swirling convergences (gyres). The so-called Great Pacific Garbage Path, off the coast of California, is the largest garbage site in the world. Plastics in this floating mass (twice the size of Texas) outnumber sea life six to one. According to the Ocean Conservancy, one million seabirds and one hundred thousand marine mammals are killed annually from plastic ingestion, and 44 percent of all seabird species, 22 percent of crustaceans, all sea turtles, and a growing list of fish species have been documented with plastic in or around their bodies.

Landfills are obviously a preferred alternative to indiscriminate illegal dumping, particularly when effective programs are in place to separate recyclable from nonrecyclable materials and the bulk of recyclables are actually recycled. Increasing numbers of landfill operations are managed to reduce problems associated with the leaching of hazardous substances into groundwater and the surrounding environment, and some landfills have installed equipment to capture methane generated by decaying materials for use as an energy source. Landfills have significant downsides, however, including operations that fall far short of best practices, health risks associated with the release of hydrogen sulfate from the decay process (linked to cancer and respiratory disease), and equity issues associated with the disproportionate location of landfills in areas populated by people of lower socioeconomic status and little or no political clout—people who are most exposed to health risks and have to contend with unpleasant odors, noisy operations, and truck congestion.

A holistic systems approach aimed at reducing material usage and eliminating waste generation altogether is advocated by materials analysts, product designers, and increasing numbers of leaders in government and industry. Termed "zero waste," "cradle-to-grave product design," or "sustainable design," the emphasis is on holistic analysis that tracks the entire life of a product from its design ("cradle") through its manufacture, distribution, use, and eventual disassembly ("grave") to enable the materials used in production to be reused or recycled rather than disposed of in landfills. Zero Waste International, an advocacy organization that provides certification and training for organizations committed to zero-waste principles, defines "zero waste" as "the conservation of all resources by means of responsible production, consumption, reuse, and recovery of all products, packaging, and materials, without burning them, and without discharges to land, water, or air that threaten the environment or human health." The aim is to change the way that materials move through society—to restructure production and distribution systems—with the goals of reducing materials requirements; facilitating product repairability, longevity, and reuse; and eliminating product waste. The three R's—reduce, reuse, recycle—are embraced as part of a larger life-cycle analysis. The approach has been adopted by the European Union and in the United States by corporations, businesses, government agencies, and institutions such as universities. Zero waste has also been characterized as a form of biomimicry: just as one organism's waste cycles through an ecosystem and provides nourishment for other life, cradle-to-grave materials circulate in closed-loop cycles to provide inputs ("nourishment") in the production of other products. In nature, nothing is wasted and everything is recycled. So should it be in human society.

3 VIEWS OF NATURE

Nature has been conceived in many ways: as a world in which spirits animate plants and animals and the places they inhabit; as a world created by God, an abode where humans are able to work on individual salvation; as an interconnected and constantly changing network of opposing yin-yang forces naturally tending toward balance and harmony; as a material order governed by physical laws and shaped over vast periods of time by evolutionary processes; as untamed wilderness that impedes civilization and must be brought under control; as a storehouse of raw materials for the production of goods that make human life more secure and enjoyable; as pristine wilderness, where harried urban dwellers are able to retreat for recreation, contemplation, and aesthetic enjoyment; as a world of things-in-themselves that cannot be apprehended save through mind-imposed (or culture-conditioned) forms of perception and understanding; and what life was like when the human imprint was far less pervasive. These are some of the images that have been entertained by people living in different cultures at various times in human history.

Many contemporary urbanites view nature as the storehouse of resources from which the necessities of life are derived: land, food, water, energy, recreational resources, etc. Science and technology are relied upon to keep nature at bay and to increase margins of safety, convenience, and comfort within human-themed environments. Excursions into the wild are enjoyed by some, but generally in small doses and in controlled circumstances. Some see "things human" and "things nonhuman" as constituting different orders of reality, in which the human order is regarded as superior to the natural order, and the natural order is regarded as having value to the extent it serves, or is made to serve, human ends. Some see human superiority as divinely ordained, while others see humankind as standing at the top of the evolutionary ladder, as life's crowning achievement.

The following takes a closer look at how nature has been conceived in a number of these traditions, including the following: nature as divine creation, as the antithesis of mind and civilization, as evolutionary product and process, as human artifact, as cultural construct, and as wilderness.

The final section (titled "the concept of nature revisited") frames a more ecological-oriented concept of nature and discusses how reconceptualization of nature shapes work in conservation biology to protect, preserve, and conserve natural resources.

NATURE AS DIVINE CREATION

Diverse views of nature are found in the Judeo-Christian tradition. Beliefs and doctrines that cast nature in a less than positive light include the view that nature is an abode over which humans have been given mastery and dominion,[1] efforts by Israelite prophets to rid religious thinking of nature gods to ensure that undivided allegiance is given to the one true God,[2] accounts that see nature as the domain of hostile energies in which the faith and fortitude of the faithful are put to test,[3] and the belief that divine cre-

1 In Genesis 1:26–29, God said, "Let us make man in our image, after our likeness; and let them have dominion over the fish of the sea, and over the birds of the air, and over the cattle, and over all the earth." In Genesis 9:1–3, God blessed Noah and his sons, and said to them, "Be fruitful and multiply, and fill the earth. The fear of you and the dread of you shall be upon every beast of the earth, and upon every bird of the air, upon everything that creeps on the ground and all the fish of the sea; into your hand they are delivered. … I give you everything."
2 The early Semites worshipped nature gods, chief among them Baal, variously characterized as the god of fertility, the protector of crops and livestock, the god of the thunderstorm, and a mighty warrior god. The Old Testament prophets saw Baal as the enemy of the God of the Israelites, and, in their attempts to extirpate these false gods, played a role in desacralizing nature.
3 For the Neoplatonist theologian, Origen, "The world of flesh is the world of demons … and gross matter the domain of Satan." For Saint Bonaventure, the created world is not wretched, but a place in which flesh and desire fetter humans and must be overcome and transcended. For Martin Luther, nature is a symbol of God's wrath and certainly not anything to admire or commune with. It is, Luther wrote, the "domain of hostile energies … which motivate the despairing soul to seek out and cling to 'the right hand of God.'" "And what

ation is hierarchically ordered, with humans above animals, animals above plants, and plants above inanimate matter.

The idea that nature is hierarchically ordered is prominent in the theology of Saint Thomas Aquinas. Based on the belief that God's creation is populated by beings variously distant from God's perfection, with humans (only humans) created in the image of God, Saint Thomas gave this summation: "We conclude, then, that lifeless beings exist for living beings, plants for animals, and the latter for man. The whole of material nature exists for man, inasmuch as he is a rational animal."[4]

If nature is cast in a negative light in these ways, one also finds positive images throughout Judeo-Christian scripture and the teachings of Catholic saints. Genesis 1:22 tells that "God saw that it [God's creation] was good." Psalms 19:1 tells how the heavens and firmament proclaim God's glory, as does Isaiah 6:3: "Holy, holy, holy is the Lord of hosts; the whole earth is full of his glory."

In Matthew 25:14–3 of the New Testament, mankind's stewardship responsibility is made explicit and emphasized: "We who are entrusted with God's property are expected to care for God's property and to distribute God's bounty justly; we will be held accountable for these stewardship responsibilities."

Saint Augustine extols values found in nature itself: "They [people who disdain nature] do not consider how admirable these things are in their own places, how excellent in their own natures, how beautifully adjusted to the rest of creation, and how much grace they contribute to the universe by their own contributions, as to a commonwealth."

Saint Francis of Assisi marveled at nature's beauty and order and expressed gratitude and respect for fellow animals and plants, all members of God's creation. In his *Canticle of the Sun*, Saint Francis proclaimed, "Be praised, my Lord, with all your creatures. … Be praised, my Lord, for our Sister Mother Earth, Who nourishes and governs us, and produces various

---

of thorns, thistles, water, fire, caterpillars, flies, fleas, and bedbugs?" he asked. "Collectively and individually, are not all of them messengers who preach to us concerning sin and God's wrath?"

4 The Aristotelian influence is unmistakable. Rationality ("rational soul") is the defining form, the essence, of human beings and is higher in the order of being to plants, which have only the power of growth ("nutritive soul") and to animals, which have, in addition, the powers of locomotion, sensation, and feeling ("sensitive soul").

fruits with many-colored flowers and herbs ... Praise and bless the Lord, and give thanks and serve him with great humility."

In the 2015 *Encyclical on the Environment* ("*Laudito Si*"), Pope Francis emphasizes the mandate of responsible stewardship pronounced in the gospel of Saint Matthew and the teachings of Saint Augustine and Saint Francis of Assisi (Pope Francis's namesake). As stated in the encyclical, "Saint Francis of Assisi reminds us that our common home is like a sister with whom we share our life and a beautiful mother who opens her arms to embrace us," adding that "This sister now cries out to us because of the harm we have inflicted on her by our irresponsible use and abuse of the goods with which God has endowed her." Also: "If we approach nature without Saint Francis' sense of awe and wonder, our demeanor will be that of master, consumers, and exploiters, unable to set limits." In another section, the encyclical states that "It is not enough to think of species merely as potential resources to be exploited, while overlooking that they have value in themselves." In regard to climate change, the 2015 encyclical makes this statement: "The problem is that we lack the vision and courage needed to confront climate change with concern for all and without prejudice toward coming generations."

Elements of these various conceptions have echoed and reechoed through the ages, and play a role in shaping the views of many people today.

NATURE AS ANTITHESIS OF MIND AND CIVILIZATION

The "modern view" is also shaped by Enlightenment ideology, articulated in its fundamentals by René Descartes and Francis Bacon. Descartes maintained that there are two fundamentally different kinds of reality: mind (*res cogitans*) and matter (*res extensa*). Bacon formulated the rudiments of the scientific method and saw the new science as an indispensable tool for controlling wild nature for the benefit of mankind. From Descartes, nature came to be seen by people living in Western societies as the dumb and machine-like "other." From Bacon, nature came to be seen as inimical to civilization and needing to be harnessed and brought under human control through scientific understanding. As these ideas were assimilated and augmented by advances in Newtonian science, nature was viewed as a machine whose operations are causally determined and capable of being fully understood—a view epitomized in this famous statement by Pierre-Simon Laplace: "An intellect, which at a certain moment would know all forces that nature set in motion, and all positions of all items of which nature is composed, ... would embrace in a single formula the movements of the greatest bodies of the universe and those of the tiniest atom; for such an intellect, nothing would be uncertain and the future just like the past would be present before its eyes."[5]

Darwinian images of nature as "red in tooth and claw" (Tennyson's phrase) were later introduced to the public mind, reinforcing the Baconian view of nature as menacing, dangerous, and the antithesis of culture.

The juxtaposition of nature with what was proclaimed to be the nobility of the human mind and human culture had the effect of further desacralizing nature. Viewed as the antithesis of "things human," nature's value was reduced to its use-value for humans, and emerging science was given the mandate to control disease, droughts, dangerous animals, and whatever else there is about nature that was seen to seen to be threatening and menacing to human life and impediments to the progress of civilization.

These themes echoed in the new America with the mandate to push back the frontier, cut down the forests for crops and pasture, build railroads and towns, and overcome the resistance of savages.

NATURE AS EVOLUTIONARY PRODUCT

Charles Darwin's *Origin of the Species*, exuberantly described by Darwin's contemporary Thomas Huxley "as the most powerful and the most comprehensive idea that has ever risen on earth," offered a new understanding of human origins. Darwin's seminal idea, grounded in his painstaking research conducted in a five-year voyage on the *Beagle* around coastlines in Africa, Asia, the South Seas, and the Galapagos of South America, challenged the idea that nature is the product of divine creation.

Darwin's hypothesis was revolutionary: all species of life have descended over time from common ancestry as the result of an ongoing process of natural selection. In his introduction to *Origin of the Species*, Darwin gave this summary account:

5 Pierre-Simon Laplace, *A Philosophical Essay on Probabilities* (New York: 1902), p. 4. Using a more recent idiom, Laplace's statement has been rendered as follows: If one knew the position, mass, and velocity of every atom in nature, one could, in principle, predict the entire future of the universe.

> As many more individuals of each species are born than can possibly survive; and, as, consequently, there is a frequently recurring struggle for existence, it follows that any being, if it varies however slightly in any manner profitable to itself, under the complex and sometimes varying conditions of life, will have a better chance of surviving, and thus, be naturally selected. From the strong principle of inheritance,[6] any selected variety will tend to propagate its new and modified form.

Darwin's theory of natural selection has since gained broad scientific acceptance and applied in many contexts. Controversial offshoots of the theory of natural selection have been advanced, notably the practice of eugenics to purify human gene pools and in schools of thought labeled as "social Darwinism." While such applications are today widely seen as aberrations of Darwin's central scientific accepts, the principle of natural selection, supplemented by advances in genetics made possible through molecular biology, finds widespread application in virtually all fields of science.

NATURE AS HUMAN ARTIFACT

The modification of landscapes, ecosystems, and geophysical systems has been occurring for tens of thousands of years as a result of human activities such as gathering, hunting, burning, cutting, grazing, planting, damming, mining, excavating, and building. Since the ramp-up of the Industrial Revolution a scant three centuries ago, human modification of nature has accelerated with mind-numbing speed. In *The End of Nature*, Bill McKibben discusses how humans have transformed not only the land and its communities of life, but also geophysical processes, including the composition of the earth's atmosphere.

Industrial and transportation technologies have wrought extensive changes on a macroscale, and genetic engineering and nanotechnology are bringing changes of comparable scope on a microscale. Other technologies are revolutionizing the manufacturing of products for use in industry, medicine, and households, including advances in robotics and 3-D printing.

6 Darwin was a contemporary of Mendel and was probably aware of Mendel's work in genetics.

These developments have led to the claim made by some commentators that nature is fast becoming a human artifact. In the most extreme formulation of the nature-as-artifact thesis, wild nature is said to have become a remade artifact engineered to satisfy human needs and enable human aspirations. Science and technology have given humans the "upper hand," including the prospect of controlling the trajectory of life on the planet through widespread application of gene-editing (CRISPR) technology. Humans are in control, and little of what was once wild nature remains.

The extreme version of the nature-as-artifact thesis is surely overstated. Human activities have extensively modified the physical environment, but there are still lands relatively undisturbed by humans, including the tundra of North America, northern Europe, and Asia, the coniferous forests south of the tundra, the Amazonian rain forest, the Tibetan Plateau, the Australian Outback, the Sahara and Gobi deserts, and the depths of the world's oceans. The International Union for the Conservation of Nature (IUCN) estimates that 45 percent of the planet's surface are "wildlands," defined as lands retaining 70 percent of their original vegetation and having a human population of less than five people per kilometer. Of the world's landmass thus classified as wildlands, approximately 15 percent is legally protected.

Clearly, also, many disturbed places have the capacity to recover from human intervention, just as life in general has the capacity to adapt to perturbations and to continue to evolve as it has for billions of years. And while technology has extended the ability of humans to protect themselves against nonhuman agents and events, nature lets us know, sometimes catastrophically, that human control is limited and tenuous. Volcanoes, earthquakes, hurricanes, and other of nature's "reality checks" temper hubris and tell us what we intuitively know anyhow—that the greater natural order establishes the boundary conditions of human thought, action, and safety.

NATURE AS CULTURAL CONSTRUCT

Rather than seeing nature as a product of human making, postmodernists emphasize that nature is a cultural construct. While this contention has numerous variations, the core assertion is that nature is how it is perceived and understood to be within a particular cultural milieu. Enlightenment thinkers saw nature as the "other": as machine-like, deterministic, inimical to civilization, and utterly devoid of intrinsic value. Romantic writers articulated very different conceptions, seeing nature as reflecting the artistry of God, as having intrinsic value, and as a sanctuary for spiritual renewal. But however nature is represented to be, the postmodernist contention is that nature is just that: a representation. Any representation is the product of cultural ways of seeing and conceiving; there is no privileged standpoint, no objective truth, no viewpoint that transcends cultural context.

Postmodernists take up the task of "deconstructing" projects and views with the objective of revealing underlying assumptions and presuppositions, hidden motives, and bias. Along these lines, postmodernists contend that ideologies are propagated by and reflect the interests of powerful elites within a culture, and that institutions and nations with relatively greater economic and political power have historically imposed their agendas on the less powerful. It is argued in this vein that the view of wilderness enshrined in the Wilderness Act of 1964—"places untrammeled by man without permanent improvement and human habitation, where man himself is a visitor who does not remain"—is one among many conceptions of nature that governments in the United States and other Western countries have sought to impose on undeveloped countries, urging them to preserve natural resources and forgo the kinds of development that have enabled the already-industrialized nations of the West and North to achieve high standards of living. Not surprisingly, critiques of this kind are also voiced by critics within the underdeveloped world.[7]

Writers in the ecofeminist tradition have joined postmodernists in rejecting ethnocentric bias in how nature is conceived, focusing more narrowly on male-oriented (androcentric) biases in inherited ideology and their perpetuation in contemporary society. Bacon's characterization of nature as female in character (as wild, dangerous, and reluctant to give up her secrets) and scientific investigation as a fundamentally male enterprise (as requiring an objective, rational, and nonemotional mind-set) have been roundly denounced by ecofeminists. Ecofeminist writers have also targeted dualisms associated with the "modern view"—mind/body, reason/emotion, objective/subjective, abstraction/embodiment, culture/nature—together with the tendency

7 A widely read commentary on Anglo-American environmentalism is Ramachandra Gua's "Radical Environmentalism and Wilderness Preservation: A Third World Critique." This widely reprinted article first appeared in *Environmental Ethics*, Vol.1, Spring 1989.

to associate superiority with male attributes (reason, objective, etc.) and inferiority with their female counterparts (body, emotion, etc.). Karen Warren characterizes such dualisms as aberrations of "value-hierarchical thinking" and the "logic of domination."[8]

Acknowledging that events, conditions, and processes in the world are disclosed to us through our perceptions of them, and that without such perceptual disclosures humans would have nothing to understand, various writers reject the conclusion that perception and understanding isolate humans utterly and completely from the world they perceive and seek to understand. Without resorting to a "naïve realism"—the view that cognition exactly and precisely discloses an independently existing world as it "really is"—Holmes Rolston articulates a "critical realism" to combat not only naïve realism but also the view that nature is, inescapably, a mental or cultural construct. With reference to scrupulously conducted scientific investigations in evolutionary biology, molecular biology, ecology, and cosmology—investigations in which explanatory hypotheses are developed and tested by observations and in which experiments are designed to confirm or deny those hypotheses—Rolston argues that, over time, we get "nature for real," not merely the cultural construction of nature. What is discovered is bound up with theoretical constructs, but the basic test of truth for scientific knowledge is exacting: a hypothesis stands only as long as its predicted results are corroborated by repeated experimentation, while non-corroborating evidence quickly discredits that hypothesis. Unlike naïve realism, critical realism recognizes the conceptual contribution of the knowing subject in the formation of theories and acknowledges the incomplete, selective, and perspectival character of scientific theories. Although no perfect correspondence exists between scientific hypotheses and reality "out there," the critical realist nevertheless maintains that rigorous science yields, or progressively yields, an accurate and reliable map of physical reality.[9]

It is misleading to erect a sharp distinction between subject and object, perceiver and perceived, culture and nature, and then to construct a reductive explanation of one in terms of the other. It is as meaningful to speak of a

8 Karen Warren, "The Power and Promise of Ecofeminism," in *Environmental Philosophy: From Animal Rights to Radical Ecology*, Michael E. Zimmerman et al., eds., (Pearson Prentice Hall, 2005).

9 See Rolston's "Nature for Real: Is Nature a Social Construct?" in *The Philosophy of the Environment*, D. J. Chappell, ed. (Edinburgh University Press, 1997), pp. 38–64.

"natural" order as it is to speak of a "cultural" order. In some places (e.g., a university or a library), cultural products and processes predominate; in others (e.g., a rainforest or a coral reef), natural products and processes predominate. Moreover, in maintaining that nature is only a cultural construct while at the same time insisting that all conceptual constructs are relative to particular cultural contexts, postmodernism is said by some writers to trap itself in a contradiction; it inconsistently privileges its own conception of nature as nonrelativist truth. As Charlene Spretnak put it, "'Postmodern' is '*most*modern' because its proponents have been socialized and educated in the scientific-humanist world view, which is dedicated to the denial of the power and presence of nature."[10]

NATURE AS WILDERNESS

The idea of nature as wilderness gained popular appeal in Europe through the writings of the European Romanticists Percy Bysse Shelley, John Keats, and Jean Jacques Rousseau, and in America through the works of the American transcendentalists Ralph Waldo Emerson, Henry David Thoreau, Walt Whitman, and John Muir—and through grand landscape paintings, photographs of nature's many moods, and the inspiring descriptions of wild places by nature writers. Nature came to be seen as magnificent places where dispirited humans living in controlled urban environments could find escape, liberation, tranquility, and inspiration. The visions and sentiments expressed in these works played a key role in changing popular attitudes toward nature as hostile and dangerous and in creating a political climate for the establishment of national parks and wilderness areas in the US and elsewhere.

While negative attitudes toward wild places have waned in contemporary western cultures, various writers have sought to debunk what they call the "wilderness myth." Critics point out that there are few places on Earth today that meet the definition of wilderness given in the Wilderness Act—that indigenous peoples occupied and modified the lands that are now designated as wilderness areas in the US and elsewhere, that attempts to set aside wilderness areas and to preserve wilderness conditions require intensive and

10 Charlene Spretnak, *The Resurgence of the Real: Body, Nature, and Place in a Hypermodern World* (Addison-Wesley, 1997), p. 66.

continuous human management of those areas, and that it dilutes responsibility for protecting areas that are less pristine. Critics have also argued that wilderness preservation as conceived in the Wilderness Act is ethnocentric and flies in the face of the basic fact that change and adaptability to change are basic dynamics in evolving ecosystems, not constancy and uniformity.[11]

There is merit in these critiques, but they, too, run the risk of blurring important distinctions and distorting facts. There are important differences, Holmes Rolston reminds us, between human-inhabited and human-uninhabited regions, between nature and culture, between the generally slow pace of biological change and the increasingly fast pace of social and technological change, and between the modifications made by indigenous peoples of the lands they occupied and those wrought by industrialized societies with bulldozers, large populations, and economies that are remarkably efficient in satisfying the voracious wants and desires of growing populations.[12]

THE CONCEPT OF NATURE RECONSIDERED

The natural world *has* been transformed to a large extent by ubiquitous human presence and by ever-expanding technological capability to exploit nature's products and processes to serve human ends, but certainly not completely so. Individual and societal views of nature *are* shaped by the beliefs, values, and institutions of the culture in which one lives, but this does not mean that "nature for real" cannot be understood through rigorous and continuing scientific inquiry, or that cultural constructs of nature that disadvantage women, people living in Third World countries, and others are immune from moral criticism and reform, or that insidious dualisms should not go unchallenged. Humans and other life *have* evolved by processes of natural selection, but the capacity of systems throughout nature to self-organize, adapt, and create novelty is also part of the story of life's evolution. Religious beliefs *do* give meaning and hope in many people's lives, but religious accounts that reject

11 Arguments for and against the "wilderness concept" are presented by the editors and contributing authors to *The Great New Wilderness Debate*, J. Baird Callicott and Michael P. Nelson, eds., (Athens, GA.: University of Georgia Press, 1998).

12 Holmes Rolston, "The Wilderness Idea Reaffirmed," in *The Environment Professional* 13 (1991), pp. 370–77.

ex cathedra evolutionary and other accounts of the relationship of humans to the broader natural order limit and distort an understanding of the human condition. None of the various conceptions of nature briefly sketched previously are adequate by themselves—or for a world in which rapidly growing human populations are transforming landscapes and consuming resources at accelerating rates, nature's regenerative capacities are being diminished, resource shortages are recurring on a global scale with ominous potential for widespread conflict and warfare, human-caused species extinctions are occurring on a scale rivaling the mass extinctions of the distant geological past, and climate change is occurring as a result of rapidly increasing concentrations of greenhouse gases in the atmosphere.

Nature is not only a product, but a vast biosphere in which life-creating and life-sustaining processes occur and have occurred for more than three billion years. The ongoing saga of life is about the capacity of many lives to adapt, survive, and thrive in their environments. It's also about the adaptive and information-based capacity of genomes (the concrete embodiment of species-kinds) to orchestrate structure and function at the cellular level, and about the capacity of ecosystems to enable a great diversity of existing kinds to coexist, coadapt, and coevolve in relations of mutual benefit and to generate new kinds of increasing diversity and complexity. Manifestations of these capacities constitute the most basic processes that occur in the planet's biosphere, and themselves are processes through which pro-life tendencies are selected in the natural order.

The wilderness concept goes awry when nature is seen as an end product that needs to preserved in a fixed, unchanging state. If the term "wild" is used to refer to processes in the natural world whose operations and products predate human life and whose continuation, while influenced by human activity, are not dependent on human life, the challenge is to preserve wildness, not wilderness. An organism, ecosystem, or bioregion is wild to the extent that its functioning is not disturbed or controlled by human beings, or to the extent that it is capable of recovering from human interference. Clearly there are relatively wild places left on Earth, and clearly many disturbed places have the capacity to recover and operate on their own.

In his essay "Walking," Thoreau said: "In wildness is the preservation of the world." The term "wildness" in this statement is sometimes interpreted as "wilderness," but Jack Turner insists that this is a misrepresentation of what Thoreau meant. Turner points out that Thoreau noted in his *Fact Book* that "wild" is the past participle of "to will" and that wildness is "self-willed land."[13] By identifying wildness with fixed conditions in nature rather than self-organizing processes that produce and sustain those conditions, we get a flawed concept of wildness. We also get, according to Turner, wilderness areas that are too small, too accommodating of human recreational interests, too controlled and regulated, and too tamed. In Turner's words: "Thus diminished, wilderness becomes a special unit of property, treated like a historic relic or ruin—a valuable remnant."[14]

In what follows, then, the term "wild" is used to refer to life and life systems whose existence is not dependent on human life and that would continue to function and adapt without human life. The concept of nature-as-wildness legislates against statements such as "Nature is what it is conceived to be" (nature as cultural construct) or "Nature is what humans make of it" (nature as human artifact), or as wilderness set-asides to be maintained in fixed states (nature as wilderness). It also mitigates against the statement that "Everything is natural." At the same time, however, it accommodates basic facts: that humans are part of nature; that culture is situated within and depends on wider natural systems; that wildness can be found in gardens and urban parks as well as set-aside wilderness areas; and that neither culture nor conditions in wild nature refer to unchanging conditions. Similar to culture, nature is both product and process and constantly undergoes change. Furthermore, it does not presuppose that nature and human culture are inevitably and irreconcilably opposed, or that the human influence on other life is necessarily good or bad. Arguments for the preservation of wilderness and wildlife are examined in subsequent chapters, but the present point is that such arguments should not be settled in advance by flawed or overly restrictive definitions of the terms "wildness," "wilderness," and "nature."

13 Jack Turner, *The Abstract Wild* (Tucson: University of Arizona Press, 1996), p. 111.
14 Ibid., pp. 84–86.

4 TOWARD AN ATTITUDE OF RESPECT

This chapter opens with a discussion of the far-reaching implications of how individuals and organizations are attitudinally oriented toward nature. After clarifying how the terms "attitude" and "attitudinal orientation" are used here, the discussion turns to a discussion of constituent elements of an attitude of respect toward nature, and to what is regarded to be essential and/or most conducive in the formation of respectful attitudinal orientation. This discussion is organized under the headings "Rekindling Felt Connections," "Telling Life's Story," "Conceptualizing Complex Biotic Systems," "Taking Stock of Nature's Manifold Value," and "Recoiling at the Destruction."

The chapter concludes with an analysis what is regarded here as four primary environmental virtues: respect, humility, gratitude, and responsibility.

ATTITUDINAL ORIENTATION

Definition of Terms

One's attitude toward anything—an election outcome, a university admissions policy, removal of a Confederate statue from a public square, collapse of a coral reef ecosystem—reflects one's affective, cognitive, and behavioral proclivities with respect to that thing: how one feels about it or relates to it emotionally; what one believes or knows about it; and how one is inclined to behave or act with respect to it.

Attitudinal orientation is conditioned by many factors, of course, including experience, the attitudes and expectations of others, the influence of parents, teachers, and role models, internalized norms, genetic predisposition, and instinct. In respect to the behavioral component of attitude, particularly in situations involving choice, intentions come into play as do constraints on choice and ability to control outcomes.

Attitudinal orientation is normative in the sense that it involves internalized norms based on what one comes to believe is good and bad and/or right and wrong. Belief-based norms incline one to act or behave in certain ways in certain situations. They are acquired over time, and also change over time as one matures and acquires experience through learning, role playing, role modeling, and how one is affected (positively and negatively) by the consequences of one's behavior. Internalized behavioral norms may not play a significant role in actual behavior, or they may come to be firmly engrained dispositions that play a strong action-directing role.

Repeated disconnects between what one believes and professes to be right (the cognitive dimension of virtue) and what one actually does (the behavioral dimension) is a sign that the normative belief is weakly held, poorly understood, and/or at odds with how one feels or is emotionally engaged in a situation at hand (the affective dimension). Such "disconnects" signal that experience (teaching, mentoring, positive and negative reinforcement, etc.) has had negligible effect in fostering the disposition to do right.

On the other hand, when the several dimensions of virtue are integrated in attitudinal orientation and one is dispositionally inclined to do right, one is judged to be a virtuous person: a person of good character who can be counted on to be truthful, trustworthy, respectful, gracious, helpful, etc.

Attitudinal Orientation Toward Nature

One's attitudinal orientation toward nature *writ large* similarly involves affective, cognitive, and behavioral components as characterized previously, but here the reference to is to fundamental conditions of life and to how one "sees" oneself in relation to—as situated in and part of—the broader natural world. Such "seeing" comprehends both beliefs and feelings in contexts involving complex sets of circumstances and influences. Behavioral proclivities and inclinations are formed in this milieu and a sense of orientation-in-the-world emerges.[1]

1 It is instructive to view attitudinal orientation as a complex self-organizing system exhibiting adaptability, creativity, open-endedness, susceptibility to exogenous shocks but at the same

A central thesis of this chapter is that an individual's or a society's attitudinal orientation toward nature has far-reaching consequences, and that elements of how many people living in urbanized and industrialized society today "see" the world is in serious need of revision. When levels of production, consumption, and waste generation exceed the capacity of natural systems to sustain those activities and, worse, degrade nature's regenerative and assimilative capacities; when the dominant economic paradigm ignores, in theory and practice, the costs associated with the diminished capacity of natural systems to perform essential ecological services; when dominant ethical traditions acknowledge the moral standing of only one species (*Homo sapiens*) among millions of other species; when environmental problems of unprecedented seriousness face people around the globe but people and governments are unwilling or unable to take corrective action; when thinking at the frontiers of contemporary science are abandoning mechanistic-reductionist models of explanation—when contemporary life is riddled with these deeply ingrained and far-reaching incongruities—it is time to reexamine fundamental questions related to the relationship of humanity to the broader natural order. It is time for attitude adjustment.

An attitudinal orientation that regards nature as the separate and distinctly different "other," as inferior and inimical to culture and civilization, and as having value only because it yields resources for human needs and aspirations, is, not surprisingly, a society in which sanctions against environmentally and ecologically destructive practices are weak. In contrast, an attitude of respect—an attitude shaped by felt connections to the land and other life; an attitude that draws intellectual support from models of explanation that go beyond mechanistic determinism to focus on relationships, interdependence, and emergent, creative change; an attitude that affirms the value of self-organizing processes throughout the biosphere and recognizes their incalculable instrumental value for human life; an attitude quickened and humbled by direct experience of nature's beauty, wonders, and power; and an attitude aroused and put on "red alert" by growing awareness of the extent to which human activities are putting human life and most other life in serious jeopardy—would, if it took hold in the minds of the general populace, be

time self-organizing capacity to recover and reestablish order and equilibrium, and emergence of synergistic effects not amenable to reductive-deterministic explanation—characteristics exhibited by all complex self-organizing systems such as organisms, genomes, ecosystems, a free market economy, and the internet.

respectful of the natural world in which we live and attitudinally inclined to put the brakes on practices and policies that are inherently unsustainable and massively harmful to nonhuman life and life systems.

In this vein, then, the following reviews what various writers have said about formative elements of an attitude of respect toward nature, organized under the general headings noted previously.

CONTRIBUTING ELEMENTS OF AN ATTITUDE OF RESPECT

Rekindling Felt Connections

An attitude of respect depends, in no small part, on the rekindling of felt connections to the land and its teaming life. Indigenous peoples everywhere have always felt and have been grounded in this connection. Writers variously contend that feelings of connectedness to the land and its communities of life initiate at a preconceptual level of experience. This awareness is variously said to be encoded in our genes, to engender innate feelings of care, to be rooted in the structure of perception and heightened when the senses are fully engaged and attuned to the sensorial field, to be associated with the unique qualities of place and the feeling of being at home, to be quickened when one is directly and attentively engaged in activities related to gardening and caring for animals, to be heightened and enriched by aesthetic experiences in nature, and to be evoked at a primordial level by the "call of the wild." World views, lifestyles, and institutional arrangements that cut one off from these elemental modes of perceiving, living, and adjusting are said to be spiritually impoverishing and to lead to a host of adverse psychological and social consequences.

Biophilia and Care

The celebrated biologist, E. O. Wilson, contends that humans have an "innate emotional affiliation to other living organisms," which he calls *biophilia* and associates with genetic predisposition.[2] Darwin viewed the feelings of kinship that animal species have with one another and we have with animals and other members of our own species as adaptive mechanisms acquired through natural

2 E. O. Wilson, *Biophilia* (Cambridge University Press, 1984), p. 31.

selection. Stephen Kellert concurs: "During the long course of human evolution living diversity was valued because of the adaptive benefits it offered us physically, emotionally, and intellectually. … Buried within the human species lies a deep and enduring urge to connect with living diversity."[3]

While the sense of connection with other life may be dimmed in the lives of urbanites cut off from direct contact with the wild, it is never far from the surface. The loss of songbirds poignantly described by Rachel Carson in her classic work, *The Silent Spring*, is a loss instantly recognized by readers because it strikes a chord deep within the human psyche. Unless conditioned to respond otherwise, people spontaneously care and are shocked when they realize that songbirds are disappearing, or when they observe cruelty to an animal, or when they see someone throwing a car battery or a large container of crankcase oil into a river or lake.

Many writers have discussed the primacy of sympathetic caring in the human psyche, variously called empathy, fellow feeling, identification, and compassion, and see such caring as the wellspring of moral response. Phenomenologist Max Scheler emphasized the need to redevelop a sympathetic capacity to decode the symbolic language of nature. "We must," he wrote, "once more look upon Nature as into the heart of a friend."[4] Albert Schweitzer sought to explain how the identification of one's own will-to-live with that of other lives throughout nature evokes a feeling of compassion capable of fostering an ethic of reverence for life.[5] Holmes Rolston calls attention to manifestations of care and caring throughout nature.[6]

Attachment to Place

Caring for the land and its life is closely related to identification with place. The German phenomenologist Martin Heidegger characterizes human being as being-in-place. For Heidegger, to live authentically is to "dwell in place" and to be "guardians of place." Inauthentic living is alienation from place, a condition that Heidegger calls "homelessness." Homelessness is an existential condition that Heidegger believes is the bane of people living in

3 Stephen R. Kellert, *The Value of Life: Biological Diversity and Human Society* (Island Press, 1996), p. 12.
4 Max Scheler, *The Nature of Sympathy* (Hamden, CT: Archon, 1978), p. 105.
5 See Schweitzer's *The Teaching of Reverence for Life* (Holt, Rinehart, and Winston, 1965).
6 See Rolston's "Care on Earth: Generating Informed Concern," in *From Physics to Metaphysics*, Paul Davies and Niels Henrick, eds. (Cambridge University Press, 2010).

modern urbanized society. People are alienated from place and experience the anxiety accompanying this alienation. Such alienation is tragic, Heidegger maintains, because people are increasingly powerless to do anything about it.

Phenomenologists Neil Evernden, Edward Casey, and David Abram emphasize that perception originates in sensory contact with the world and is essentially participation in the world. "It is misleading," Evernden writes, "to speak of an isolated self surveying a world, for the person is from the start *in* the world, and consciousness is *of* the world."[7] For Casey, place is where one's sensing body integrates with the environment in which it finds itself situated. "Far from being an inert support for egoic mind," Casey writes, "the lived body takes us out of our skin into the world's flesh, back into *place*."[8] It is through sensation that we come into direct contact with the place we inhabit, and it is through sensation that the sensible and the sentient interact in mutual disclosure.

Casey's student, David Abram, recounts how direct sensuous contact and intimate relations with the land is reflected and preserved in the oral stories and songs of indigenous peoples and how, for oral peoples, language "arose not only as a means of attunement between persons, but also between persons and the animate landscape." He discusses how creation of the alphabet gave humans the ability to interact with written signs in abstraction from the earthly things they represent and how electronic-based communications have evolved into a vast cognitive realm in which "the senses—once the crucial site of our engagement with animate earth—become mere adjuncts of an isolated and abstract mind bent on overcoming an organic reality that now seems disturbingly aloof and arbitrary." Rather than dwelling in place, we view ourselves as occupying a set of coordinates in space. When we say that we live in the US, or British Columbia, or New Mexico, "we say very little about that earthly place we inhabit, but simply establish our temporary location within a shifting matrix of political, economic, and civilizational forces struggling to maintain themselves." The danger, Abrams believes, is that "we may come to believe that our breathing bodies really inhabit these abstractions, and that we will lend our lives more to consolidating, defending, or bewailing the fate of these ephemeral entities than to nurturing and

7 Neil Evernden, *The Natural Alien: Humankind and Environment* (University of Toronto Press, 1985), p. 59.

8 Edward Casey, *Getting Back Into Place: Toward a Renewed Understanding of the Place World* (Indiana University Press, 1993), p. 262.

defending the actual places that physically sustain us." The basic need, Abram believes, is to ground ourselves, once again, in the links between the abstract world and the perceptual terrain that surrounds us. "For it is only at the scale of our direct, sensory interactions with the land around us that we can appropriately notice and respond to the immediate needs of the living world."

Dwelling in place is characterized by Edward Relph as a homeward-directed sentiment, one that is comfortable, detailed, diverse, and ambiguous without confusion. To experience this sentiment is to be an "insider": to identify with place, to feel care, concern, and warmth for it, to feel that one belongs there. For Relph, the sense of being "at home" is acquired through community associations, and is an orientation that individuals experience to a greater or lesser degree.

Barry Lopez, too, reminds us that intimate understanding of physical place comes from attending to it with all the senses and with an attitude unaffected by assumptions, expectations, and guile. We must, he says, "Put aside the bird book, the analytic frame of mind, a compulsion to identify and sit still."[9] "The key," he adds, "is to become vulnerable to place. If you open yourself up, you can build intimacy. Out of such intimacy you may come to a sense of belonging."[10] Further: "The land retains an identity of its own, still deeper and more subtle than we can know."[11]

It is reported that a Native American elder made this comment about settlers in the American Midwest: "Looks like they are finally beginning to learn how to belong here."

Mark Sagoff also champions the concept of place in environmental discourse: "It brings together human, environmental, and natural history," he writes, "and is particularly valuable in helping us to understand what we deplore about the human subversion of nature and what we fear about the destruction of the environment."[12] "Unlike the concept of space," Sagoff says, "place refers to land whose significance is not summed up by the consumer

9 Barry Lopez, "A Literature of Place," in *The University of Portland Magazine*, Summer 1997, p. 25.

10 Ibid., p. 25.

11 Barry Lopez, *Arctic Dreams: Imagination and Desire in a Northern Landscape* (London: Pan, 1987), p. 228.

12 Mark Sagoff, "Environmental Ethics: An Epitaph," in *Resources*, Spring 1993, p. 6. See also Sagoff's "Settling America: The Concept of Place in Environmental Ethics," in *Journal of Energy, Natural Resources and Environmental Law*, 12, pp. 351–418.

preferences of individuals; land is far more than delineated space, more than property, more than a commodity."

A large body of nature writing is devoted to the description of the unique features and charms of particular places. Thoreau's observations at Walden Pond and John Muir's colorful descriptions of the High Sierras are classical examples of this large genre of writing.[13]

Gardening and Taking Care of Animals

The domestication of animals and the cultivation of crops for human survival, convenience, and enjoyment have a long history and have played an important role in shaping attitudes toward nature. People who work the land, have pets, and care for farm animals know first-hand the necessity of paying attention to the needs of the land, plants, and animals with which they directly interact and assume custodial responsibilities.

Michael Pollan has written about what he has learned from years of effort as a gardener.[14] Perhaps the most basic lesson is that successful gardening depends on working in harmony with nature's ways. "The trick is to put her knowledge to our purpose in the garden." Based on personal and sometimes frustrating experiences, Pollan articulates some of the insights of a "garden ethic." This ethic is avowedly an enlightened egoism, but one that shows respect for the wildness that one encounters "right there" in one's soil and plants. The gardener's respect for nature is not abstract and not romantic ("He's seen her ruin his plans too many times for that"), and he feels he has a legitimate quarrel with her weeds, wounds, storms, plagues, rot, and death. He recognizes at the same time, however, that it is a mistake, indeed an impossibility, to reach for outright victory. He does not see nature as a neutral, fixed backdrop but as alive and changing all the time in response to innumerable contingencies, including his own presence. He does not see himself as outside nature, and he doesn't think that his interactions with nature are inherently good or bad. He believes it is possible to make distinctions between kinds and degrees of intervention. He looks for local answers and relies on his skill to make and apply informed distinctions. He has learned

13 A history of this genre is provided by Thomas J. Lyon in his *This Incomparable Land: A Guide to American Nature Writing*. See also *The Norton Book of Nature Writing*, edited by John Elder (W. W. Norton, 1980).

14 Michael Pollan, *Second Nature: A Gardener's Education* (Grove Press, NY, 1991).

that nature is, above all, a pragmatist, and that the good gardener borrows his methods, and sometimes his goals, from nature herself.

Farmers, ranchers, landscapers, gardeners, and others who interact on a daily basis with the land, its soils, and its nonhuman inhabitants do not see themselves as outside nature. They look to nature for guidance, and they go about their work with respect for the nonhuman entities with whom they interact.

Aesthetic Encounters: Awe, Beauty, and Wonder

Encounters in the wild have a powerful effect on the human psyche and give rise to aesthetic responses described by many writers. To experience the awesome power of nature, to delight in its beauty, or to find oneself absorbed in wonder about its mysteries and remarkable productions is to find oneself profoundly affected.

Awe and the Sublime

Edmund Burke characterized the sublime in the mid-eighteenth century as an emotion produced in the human mind by what he called "terrible objects," objects whose magnitude or power is so great that they evoke a sense of awe and a peculiar kind of delight. Provided that the terrible "does not press too close"—is not too threatening, too imminent, too dangerous—the encounter produces "the strongest emotion which the mind is capable of feeling."[15] This powerful and strangely pleasing emotion is not sheer terror because it is accompanied by the realization that one is a spectator, not a victim, of the terrible. Once Burke, Immanuel Kant, and other late eighteenth-century aestheticians had deciphered this emotion and introduced it to the educated public, it quickly became the rage of Romantic Age adventurers, poets, nature writers, and landscape artists. In 1785, Jean-Jacques Rousseau said, "I must have torrents, rocks, pines, dead forest, mountains, rugged paths to go up and down, precipices beside me that frighten me, for the odd thing about my liking precipitous places is that they make me giddy, and I like this giddiness greatly, provided that I am safely placed." Ruskin, Keats, and Shelley expressed similar sentiments in England as did Emerson, Thoreau, and Muir in America. When Keats first beheld the Alps in 1816, he was

15 Edmund Burke, *A Philosophical Enquiry into the Origin of our Ideas of the Sublime and the Beautiful*, Adam Phillips, ed. (Oxford: World's Classics, 1990).

overwhelmed: "I never knew, I never imagined what mountains were before. The immensity of these aerial summits excited, when they suddenly burst upon the sight, a sentiment of ecstatic wonder, not unallied to madness."[16]

The experience of the sublime produces a sense of awe: one realizes that one is in the presence of forces that transcend the mundane and that one's power and place in a larger scheme of things are limited. Diverse encounters produce this felt realization: standing before a range of soaring mountain peaks, being on a lake or ocean during a storm, beholding the countless millions of stars at night, or gazing down a thousand-foot precipice. In one of his most remarkable experiences, John Muir described a fierce spring storm in the High Sierras. With the wind howling, he chose a sturdy Douglas fir, climbed to the top, lashed himself to its trunk, and reveled as he swung to-and-fro in the gale. Muir describes how this intense experience was followed by a peaceful calm: "When the storm began to abate, I dismounted and sauntered down through the calming woods. The storm-tones died away, and, turning toward the east, I beheld the countless hosts of the forests hushed and tranquil, towering one above another on the slopes of the hills like a devout audience. The setting sun filled them with amber light, and seemed to say, while they listened, 'My peace I give unto you.'"[17]

Experience of the awesome is attitude-adjusting. One may be terrified, delighted, exhilarated, or humbled by exposure to nature's forces and ever-changing moods, but underlying the emotional response is the realization that one is in the presence of transcendent forces. Confronted with the awesome, ego is scaled back and humility sets in.

Beauty

Mountains, deserts, grassland prairies, rainforests, oceans, rivers, lakes, coral reefs, and many other places in wild nature are places where one can also experience another many-faceted aesthetic emotion: beauty. The beauty of these places is a powerful attraction for people seeking solace or escape from the frenzy of urban life, and for painters, poets, and writers motivated to communicate nature's beauty to others.

---

16 Percy Bysshe Shelley, *Peacock's Memoirs of Shelley: with Shelley's Letters to Peacock*, H. F. B. Brett-Smith, ed. (London: H. Frowde, 1909).
17 John Muir, *The Mountains of California*, cited in Ann Ronald, *Words for the Wild* (Sierra Club Books, 1987), p. 121.

Robert MacFarlane comments that there is nothing quite as enchanting as a mountain sunset or sunrise, a meadow carpeted with wildflowers, or the view one gains of canyons, forests, rivers, and valleys from a mountaintop. The beauty of mountains is not competitive, MacFarlane observes, but contemplative and revelatory. Mountains present grand panoramas, offer new perspectives, and enable one to witness an infinite variety of forms, colors, textures, and shades of light. He notes that the plurality of light has probably attracted more comment than any other dimension of mountain beauty. When sunshine is punctuated by falling snowflakes, mountain light flickers and dazzles. During a thunderstorm it flashes with ostentatious splendor, and then may radiate as the most vivid of rainbows. It brightens and dims as shadows are cast by moving clouds. It radiates off snow and ice fields with such intensity on clear days that one cannot look directly without risking snow blindness. When it shines from behind rock ridges and spires, it creates stunning architectural effects. When it shines through stands of aspen or pine, it has a glimmering, impressionist quality. It takes on a tranquil matt-like quality at dusk, unifying the whole landscape with a single texture. And then there is Midas light, the rich yellow light that spills lengthways across the terrain, turning everything it touches to gold.[18]

Wonder

Encounters in nature also ignite a sense of wonder. Witnessing the ridges and scoring in rock where a glacier has passed, feeling the soft pelts of moss on rocks and on the lee side of boulders and trees, watching the swirls of water beneath a plate of ice, observing the unique architecture and ephemeral fragility of a snowflake, watching the rapid buildup of nearby storm clouds, hearing a hillside come alive with moving water after a rain shower, watching and being watched by a mother bird protecting her eggs—none of these are trivial experiences. They are the kinds of experiences that inspired Muir, Thoreau, and many other nature writers, and it is their detailed and, at the same time, rapturous accounts that have evoked wonder and rapture in their readers.

---

18 Robert MacFarlane, *Mountains of the Mind* (New York: Vintage Books, division of Random House, 2004). MacFarlane provides poignant descriptions of the beauty and allure of mountains and, more generally, discusses the history, psychology, and aesthetics of mountaineering.

Nature's creations also ignite wonder about events that unfold in deep time: how the starry heavens came to be and what are their destinies, how continents were formed and reformed, or how fossil and mineral deposits laid down in ocean beds have been transported to mountaintops. Scientific investigation is a rigorously conducted enterprise, but it is one that initiates in wonder and amazement.

Call of the Wild

The last lines of Gary Snyder's poem, "The Call of the Wild,"[19] evoke the realization that the silencing of the last coyote's call is a loss of connection with wildness not only in the world but in our souls as well. The howl of a coyote, the melody of a songbird, the chorus of voices emanating from a pond at nightfall are becoming less and less audible against the background roar of city life.

Although many people living in urban environments are losing touch with wildness, it is never far from the surface. In the words of Jack Turner, "The easiest way to experience a bit of what the wild was like is to go into a great forest at night alone. Sit quietly for a while. Something very old will return."[20] Voices in nature rekindle vestiges of wildness in our own psyches, although those voices are dimming.

Telling Life's Story

An attitude of respect comes from a rekindling of felt connections—elemental caring, attachment to place, gardening and caring for animals, aesthetic encounters in nature, and tuning in to the beckoning call of the wild—but, no less, it comes from the telling of life's story. In the words of Stephen Gould, the story of life is the "greatest story ever told."[21] It is the story of the immensity of evolutionary time and the great geophysical forces that have changed climates and have repeatedly reshaped continents and oceans; of the hundreds of millions of species that have flourished, sometimes in great bursts of creativity, and then have perished, sometimes in epic mass

---

19 Gary Snyder, *Turtle Island* (New Dimensions, 1974), p. 21.

20 Jack Turner, *The Abstract Wild*, Op., cit., p. 26

21 Stephen Jay Gould, ed., *The Book of Life* (W. W. Norton, 2001). Chapters of *The Book of Life* are written by contributing scientists, including Michael Benton and John Sepkoski cited in the following notes.

extinctions; of life's countless adaptations and innovations and the increasing complexity of evolving life forms; of the many paths that have been blazed through the profusely branching Tree of Life, including the branch that has led to "man the wise."

Holmes Rolston, Thomas Berry, and others emphasize the importance of telling life's story and its many sub-narratives, for it is only in so doing that humans will begin to understand their place in the greater natural order. "I cannot," Rolston says, "give you an argument explaining how humans arrived, some logic by which the Earth story eventuates in *Homo sapiens*. ... What I can do is invite you as a historical subject to appreciate the story that lies in, with and under the Earth we inhabit, to enrich the story by telling it. You can be a microcosm of the macrocosm and enjoy your storied residence here."[22]

Many who contemplate the enormity of life's journey speak of how they are filled with a sense of awe, of how they feel that they are encountering something sacred and are participants in a process that is profoundly significant. Chaos-complexity theorist Stuart Kauffman is not hesitant to speak in such terms: "We latter-day players are inheritors of almost 4 billion years of biological unfolding. If profound participation in such a process is not worthy of respect, if it is not sacred, then what might be?"[23]

Thomas Berry offers this perspective: "Within the solar system, the earth is the immediate context of human existence. ... Beyond the sun, however, is our own galaxy, and beyond that is the universal galactic system that emerged some 15 billion years ago through some ineffable mystery. ... We suppose that the universe itself is *the* enduring reality and *the* enduring value even while it finds expression in a continuing sequence of transformations. In creating the planet Earth, its living forms, and its human intelligence, the universe has found, so far as we know, the most elaborate manifestation of its deepest mystery. Here, in its human form, the universe is able to reflect on and celebrate itself in a unique mode of conscious self-awareness."[24] Berry calls this the "miracle of mankind." In human beings, the cosmos becomes

22 Holmes Rolston, "The Human Standing in Nature: Storied Fitness in the Moral Overseer," in *Values and Moral Standing*, W. Summer, D. Dallet, and T. Attig, eds. (Bowling Green University Press, 1986), p. 97.

23 Stuart Kauffman, *At Home in the Universe: The Search for the Laws of Self-Organization and Complexity* (Oxford University Press, 1995), p. 303.

24 Thomas Berry, "The Viable Human," in *Environmental Philosophy: From Animal Rights to Radical Ecology*, Michael Zimmerman et al. eds. (Prentice-Hall, 2001), p. 175.

aware of itself, is able to contemplate itself, and, in this awareness and contemplation, becomes responsible for itself.

The immensity of what various writers call "deep time" puts humanity's aspirations and travails in a needed perspective. The time scale of evolving life is depicted by Michael Benton in this way: "In a book of 100,000 words, if life began around 4 billion years ago, each word covers 40,000 years and each letter a period longer than the whole of recorded human history."[25]

Another portrayal imagines that life on Earth is compressed into a movie that begins on January 1 and lasts one year. In this movie, the debut of humans would occur late on the last day of the year, about seventeen minutes before the movie ends. The appearance of human agricultural communities would occur with about eighty-five seconds left, and the onset of the Industrial Revolution with about three seconds left. If the original script had been written at the outset of the Industrial Revolution but a producer of the movie today decided to add events that transpired between 1750 and the present, the movie's final fraction of a second would have been astonishing to a person living in 1750, so sweeping have been the changes wrought by human population growth, urbanization, and economic growth. Table 1-1 depicts the emergence of life forms in evolutionary time and the approximate time each life form would have made its debut in the imaginary one-year movie.

Contemplation of momentous events in life's history and the unfolding of these events in evolutionary time dispel the notion that life is all about humans. Humans are no more at the center of life's history than Earth is the center of our solar system, or our solar system is the center of galaxies that extend beyond. *Homo sapiens* are unique and special in ways that merit special recognition and celebration, but our species is a latecomer and its fate in the continuing unfolding of life's history—measured in geological ages and eons, not centuries and millennia—is hardly certain. Cataclysmic events in the distant past have led to extinctions on a mass scale, and once dominant species have disappeared from the scene. Scientists have found evidence that a super volcano in Indonesia about seventy thousand years ago nearly eliminated *Homo sapiens* (survivors in Africa are estimated to have been as few as three thousand individuals), and have recounted many other examples of how humans have been the hapless victims of volcanoes, earthquakes,

25 Michael Benton, *The Book of Life*, Op. cit., p. 110.

Table 1-1 Life's Evolution Depicted in a One-Year Time Frame

EVOLUTIONARY MILESTONES	TIME OF FIRST APPEARANCE (YEARS AGO)	EQUIVALENT ANNUAL CALENDAR TIME	PLANETARY PRESENCE AS PERCENTAGE OF AGE OF EARTH
Bacteria	4.0 billion	February 16	87.0%
Nucleated Cells	1.4 billion	September 10	30.5%
Multicellular Organisms	700 million	November 5	15.2%
Land Plants	435 million	November 26	9.5%
Insects and Reptiles	350 million	December 3	7.6%
Mammals	225 million	December 13	4.9%
Primates	60 million	December 26	1.3%
Humans	150 thousand	11:42:51 p.m., 12/31	0.003%
Agricultural Communities	12 thousand	11:58:37 p.m., 12/31	0.00003%
Industrial Revolution	3 hundred	11:59:57 p.m., 12/31	0.000007%

hurricanes, climate change, and other of nature's powerful forces since the dawn of civilization.

Reconstruction of Darwin's Tree of Life based on comparisons of molecular sequences in genes that code for ribosomal RNA exhibit three main limbs: bacteria, archaea, and eucharya, with plants and animals forming only small twigs at the top of one branch of the eucharya limb.[26] Bacteria and archaea are microscopic organisms with prokaryotic cell organization (their genomes consist of a single strand of DNA on a single chromosome unbounded by a nuclear membrane) and harvest energy in a wide range of ways from chemical reactions. Plants and animals exhibit eukaryotic cell organization (their genomes reside in 2–600 chromosomes, each in a combination of DNA, RNA, and protein within a membrane-bounded nucleus) and derive energy through anaerobic photosynthesis (in the case of plants) and aerobic respiration (in the case of animals).

26 Andrew Knoll, *Life on a Young Planet: The First Three Billion Years of Evolution on Earth* (Princeton University Press, 2003), p. 24. In regard to the contributions made by molecular biology to our understanding of life, Knoll made this comment: "Experts argue about its details, but all biologists agree that our ability to draw Darwin's great Tree of Life in its entirety constitutes one of the great intellectual achievements of the late twentieth century."

The story of life is usually told as stories about humans, dinosaurs, and megafauna, but these are certainly not the only storylines in a much larger narrative. One central narrative recounts how mitochondria came to be incorporated in the cell structure of an early eukariote[27] and how this led to the development of plastids and anerobic photosynthesis in plants and aerobic respiration in animals. As first hypothesized by cell biologist Lynn Margulis in 1967 and confirmed by later molecular biology, neither plastids nor mitochondria are produced directly by the eukaryotic cell itself; rather, they divide by simple fission controlled by their own DNA, not the DNA of the host cell.

Eukaryotic cell structure led to the emergence of organisms with multiple cells and this, in turn, led to the prolific diversification of plant and animal life. Bacteria often have many-celled bodies, but each cell is almost identical. Only eukaryotes can variegate their cell types into skin, bone, muscle, blood, leaf, bark, and seed, and these into diverse shapes, tissues, and functions."[28]

We humans not only have ancient ancestors, but are constituted, literally, of nonhuman life. Mitochondria are contained in the trillions of cells that make up the human body, and it is mitochondria that supply energy in all of these cells—just as they have in the cells of all animals for over a billion years. The story of mitochondria is a striking illustration of endosymbiosis. Another illustration is the symbiosis between humans and the more than five hundred species of microbes that reside on our tongue and eyeballs, in our ears and nose, all over our skin, and in our gut. As Jeffrey Gordon put

---

27 The remarkable incorporation of mitochondria in the cell structure of a eukaryotic organism must have occurred around 1.8 billion years ago, sometime after the "oxygen revolution" since mitochondria are oxygen users. In the *Book of Life*, Op. cit., contributing scientist John Sepkoski recounts how the "oxygen revolution was launched by a strain of bacteria (cyanobacteria) that evolved the ability to extract hydrogen from water rather than hydrogen sulfide, to combine this hydrogen with carbon dioxide to form carbohydrates, and to transport the waste oxygen molecules outside their cells before damage was done." As Sepkoski goes on to explain: "The increasing populations of cyanobacteria on the sunlit bottoms of the new shallow seas relentlessly pumped out free oxygen, and the Earth rusted. From about 2.2 billion to 1.8 billion years ago, huge volumes of banded iron formations … accumulated in the oceans, rusted, and precipitated to the bottom. Once the reduced iron ran out, there was no other chemical sink big enough to hold the continuing biological surge of oxygen. The buildup of this gas dissolved in water then escaped into the atmosphere, and with growing concentrations of oxygen, pink skies turned blue, brown seas became azure, and the fumes of methane and hydrogen sulfide given by simple life forms were swept away, thus setting the stage for the emergence of eukarya and the vast profusion of life that followed."

28 John Sepkoski, *The Book of Life*, *Op. cit.*, pp. 44–45.

it, "This is a strategic alliance, a symbiosis between mammals and microbes that goes back millions of years. … We're really a composite of species. We have human cells, but we have ten times more microbial cells."[29] The host of organisms within and on our bodies are not invaders, but organisms that have co-evolved with animals over geological ages, working together to advance common causes. Gut microbes, for example, are essential for the digestion of food and the production of vitamins and are defenders against disease.[30]

The omnipresence of bacteria is well known, but archaea, too, are widely distributed across the earth and its oceans. Some archaea live in very unusual places: in very hot hydrothermal vents at the bottoms of seas, in waters ten times saltier than the ocean, and in highly acidic mine waters. It is fortunate for plants and animals that microscopic organisms are robust survivors[31] and perform simple tasks well, because eukaryotic life in general would not be possible otherwise. As Andrew Knoll recounts, "The cycles of carbon, nitrogen, sulfur, and other elements are linked together into a complex system that controls the biological pulse of the planet, and it is the metabolic pathways of prokaryotes that sustain these cycles. … Prokaryotic metabolism forms the fundamental ecological circuitry of life. Bacteria, not mammals, underpin the efficient and long-term functioning of the biosphere."[32]

Conceptualizing Complex Biotic Systems

On the mechanistic model, a whole is an aggregation of discrete parts that act upon each according to deterministic causal laws. Likening a whole to a machine, this view holds that there is nothing to be known about a whole than cannot be known by complete knowledge of its constituent parts and their causally determined relationships. A whole just is the sum of its

29 Jeffrey Gordon, "Growing on You" in *National Geographic*, November 2005.

30 Another notable example of endosymbiosis is the association of algae and fungi to produce a single successful organism, where the algal component provides photosynthetic ability and the fungal component contributes structure and nutrient uptake. Another is the harboring of unicellular algae within the tissue of a reef coral, an arrangement in which each derives nutrients from the other.

31 Many prokaryotic organisms have displayed a remarkable resistance to extinction, especially the blue-green algae, which have persisted for billions of years with little change. This resistance is attributable to a variety of factors, including immense population sizes and the ability to reproduce quickly, to track and adapt to changing environments with ease, and to persist in a dormant stage when conditions are unfavorable to growth.

32 Knoll, *Life on a Young Planet*, Op. cit., p. 23.

interacting parts and nothing more. The sweeping implications of this model of explanation were put by Pierre Simon de Laplace in this way: "We may regard the present state of the universe as the effect of the past and the cause of the future. An intellect which at any given moment knew all of the forces that animate nature and the mutual positions of the beings that compose it, if this intellect were vast enough to submit the data to analysis, could condense into a single formula the movement of the greatest bodies of the universe and that of the lightest atom; for such an intellect nothing could be uncertain and the future just like the past would be present before its eyes."

The mechanistic-deterministic model is unwaveringly reductionist. Phenomena occurring within a complex system can, in principle, be completely explained in terms of phenomena occurring in constituent, less complex subsystems. Accordingly, phenomena in biological systems (studied in biology) can, in principle, be fully explained by the explanation of phenomena in chemical systems (studied in chemistry), just as phenomena in chemical systems can, in principle, be fully explained in terms of phenomena in physical systems (studied in physics). Mental phenomena are similarly accounted for in reductionist terms. According to the mind-brain identity theory, a mental event (e.g., a thought, a memory, a feeling) is nothing more than a particular pattern of neural processes in a human brain. A complete understanding of brain activity would yield all there is to know about the life of the mind.

Contemporary science is telling a different story: the behavior of subatomic particles does not reveal definite movement or specific location, the theoretical possibility of predictive certainty has been abandoned, and systems are thought to generate unique phenomena that are not amenable to reductive explanation. Electrons act together in strong magnetic fields to form new types of particles in atomic structures. Atomic structures and the properties of bulk matter emerge from increasing levels of complexity in physical systems. Chemical autocatalytic processes allow living systems to self-replicate, to incorporate self-corrective feedback mechanisms, and to adapt to changes in their environments. Evolution creates increasingly complex systems and, with this complexity, novel phenomena and new patterns of relationships. Increasingly complex levels of inner organization, information processing, and awareness of self and environment have emerged and continue to emerge.

Thomas Berry speaks of the need to move beyond mechanistic-deterministic-reductionist explanation in this way: "Our observational sciences presently have moved beyond the mechanistic understanding of a so-called objective world as it was known in the past few centuries of Newtonian physics. If formerly we knew by downward reduction processes that considered the particle as the reality and the wholes as derivative, we now recognize that it is even more important that we integrate upward. ... To understand atoms, we must see these elements in their central role in molecules, megamolecules, in cellular life, organic life, even in intellectual perception."[33]

Attention has shifted to models of explanation capable of accounting for the capacity of complex systems to adapt, create new relationships, reestablish internal order when exogenous shocks cause disorder and systemic dissolution, and to generate qualitatively unique phenomena.

It is instructive in this regard to recount developments in the field of chaos-complexity theory, heralded by many writers as being as important as relativity theory and quantum mechanics. Henri Poincare (1854–1912) is regarded as a seminal figure in the development of chaos-complexity theory, primarily because he demonstrated that Newton's linear mathematics could not yield predictive certainty when three bodies (earth, moon, and sun) acted upon one another. Using nonlinear equations, Poincare showed that small differences in initial conditions produce unpredictable differences in equation solutions. About eighty years later, Edward Lorenz arrived at a similar conclusion in connection with his study of weather systems; using differential equations he demonstrated that small changes in initial conditions can lead to unpredictable changes in subsequent system behavior. Graphed solutions to the model's three equations revealed a fractal shape that resembled a butterfly. Lorenz called this fractal a "strange attractor" and emphasized how it appears to act on the system as a whole, functioning as a "collector of trajectories of perturbation." The work of another pioneer, Ilya Prigogine, showed that far-from-equilibrium conditions in dissipative systems lead to behavior very different from what would be expected from the second law of thermodynamics; rather than entropic degradation and chaotic dissipation, order and more complex structures sometimes emerge from simpler ones. Prigogine observed that the more complex a system is, the more vulnerable it is to perturbations and the more energy it requires to

33 Thomas Berry, *The Great Work* (New York: Bell Tower, 1999), p. 24.

maintain stability and coherence. When a complex system becomes highly unstable (aperiodic), small disturbances can lead to abrupt and unpredictable changes. But, Prigogine adds, "We know now that nonequilibrium, the flow of matter and energy, may be a source of order."[34] With Prigogine, the focus of investigation in chaos theory shifted to an understanding of the dynamics of complex systems displaying the capacity for self-organization. On still another front, Mitchell Feigenbaum formulated ratios that describe the critical state at which a system either dissipates or transitions from chaos to order.[35] These ratios are used to predict the onset of turbulence in systems of many kinds.

A rich vocabulary has evolved to characterize the capacity of complex systems to self-organize and continuously renew themselves, including self-production (autopoiesis), self-reflexivity (autognosis), and self-regulation (autonomics). Self-organization is variously characterized as a phenomenon involving energy and information exchange, patterning, networking, coupling, synchronization, synergy, and conservation of what works.

As recounted by Michael Waldrop, systems such as a cell, a genome, an ecosystem, a free market economy, and the internet are all *complex* ("Think of the quadrillion of chemically reacting proteins, lipids, and nucleic acids that make up a living cell"), *spontaneously self-organizing* ("The genes in a developing embryo organize themselves in one way to make a liver cell and in another way to make a muscle cell"), *adaptive* ("actively turning whatever happens to their advantage"), and *dynamic* ("making them qualitatively different from static objects such as computer chips and snowflakes"). "At the same time, however, their peculiar dynamics is a far cry from the weirdly unpredictable gyrations known as chaos, [which] by itself doesn't explain the structure, the coherence, the self-organizing cohesiveness of a complex system."[36]

Stuart Kauffman, an interpreter of chaos-complexity theory in evolutionary biology, sees more than selectivity for survival and random variation in

34 Ilya Prigogine and Isabele Stengers, *Order out of Chaos: Man's New Dialogue with Nature* (Bantam Books, 1984), p. xxvii.

35 Feigenbaum's work built on the study of noise interference by Benoit Mandelbrot in the 1970s. Mandelbrot discovered distinct ratios between order and disorder on any scale he used. Geometrical graphs of these ratios revealed a self-similar fractal pattern.

36 M. Mitchell Waldrop, *Complexity: The Emerging Science at the Edge of Order and Chaos* (Simon & Schuster, 1992).

evolutionary ecosystems. Kaufmann's hypothesis is that self-organization in complex systems is a necessary condition of evolvability. He cites the bilipid cell membrane and the genomic network orchestrating cellular activity as self-organizing systems that are robust enough to be molded and further refined by processes of natural selection. Quoting Kauffmann: "The cell membrane is a bilipid membrane, stable for almost 4 billion years both because it is robust and ... readily malleable by natural selection. The genomic network lies in the ordered regime, perhaps near the edge of chaos, because such networks are readily formed, but also because such systems are structurally and dynamically stable, so they adapt on correlated landscapes and are able to be molded for further tasks."[37] Self-organization in complex systems is also the key, Kaufmann believes, to explaining the reappearance of millions of species following mass extinctions such as the Cambrian and Permian extinctions.

"Evolution left us stuck on earth with no ladder to climb," Kauffman wistfully notes. "Random variation, selection-shifting. Here lies the brooding sense of accident, of historical contingency, of design by elimination."[38] But Kauffman sees more to it than this: "Evolution is surely 'chance caught on the wing,' but it is also the expression of underlying order." This realization enables us feel "at home in the universe."[39]

On a planetary scale, James Lovelock hypothesized that Earth is an entity of great complexity that regulates the mix and concentration of gases in the atmosphere in ways that make life possible. After the name given by the ancient Greeks to goddess Earth, Lovelock called this hypothesis the Gaia Hypothesis. In his words, "The entire range of living matter on Earth, from whales to viruses, and from oaks to algae, could be regarded as constituting a single living entity, capable of manipulating the Earth's atmosphere to suit its overall needs and endowed with faculties and powers far beyond those of its constituent parts."[40]

---

37 Stuart A. Kauffman, *At Home in the Universe: The Search for the Laws of Self-Organization and Complexity* (Oxford University Press, 1995), p. 188.

38 Ibid., p. 7.

39 Ibid., p. 189.

40 James Lovelock, *Gaia: A New Look at Life on Earth* (Oxford University Press, 1987), p. 9. Lovelock in some passages refers to Earth as an "organism." Critics suggest that the metaphor of a very complex ecosystem (rather than a giant organism) would be a more apt choice to characterize Earth's self-regulating pro-life tendencies.

Taking Stock of Nature's Value

Full appreciation of nature's value involves the recognition of inherent values *in* nature, not only the recognition of value *of* nature for human beings. If one believes that nature has value only because it is *of* value to humans, one's sense of respect for nature is truncated; one does not respect things judged to be valueless in their own right.

Inherent Value in Nature

As will be discussed in the next chapter, an anthropocentric account of value holds that values originate in human experience, and that valuers and values are associated exclusively with human life and culture. On this account, something is not a value unless it is judged by a human to be a value. Distinctions are made between intrinsic value (things valued in and of themselves, or things valued for their own sakes) and extrinsic value (things valued because they are instrumental in bringing about other values), but the anthropocentric claim is the same: to say that something has intrinsic or extrinsic value is to say that some human being judges that it has intrinsic or extrinsic value. For an anthropocentric value theorist, a world without humans would be a world without value.

In the previous chapter, value was associated with three of the most fundamental processes or capacities that occur in the biosphere: the capacity of organisms to realize biological over their lifetimes in challenging environments (to self-actualize); the capacity of genomes to orchestrate the complex patterns of cell differentiation and function that, over time, make an organism an integrated structural whole, thus making self-actualization in an organism and species perpetuation possible; and the capacity of ecosystems to enable many individuals of many genomic kinds to coexist, coadapt, and coevolve in relations of mutual support and benefit. New value terms are coined to characterize such value (e.g., systemic or constitutive value). Self-actualization constitutes the good of individual living beings; genomic expression constitutes the good of species-kinds; and coadaptation in relations of mutual advantage constitutes the good of ecosystems. Self-actualization at the level of the organism, genomic expression at the level of the species, and coadaptive fitness at the level of the ecosystem are themselves valuational processes occurring in self-organizing systems (are systemic values).

Using the language of chaos-complexity theory, it is meaningful to think of values as organizing trajectories of activity. When values provide meaning and direction in an individual's life, things tend to go well for that individual. Similarly, when shared values in organizations maintain an attractive force and give direction to collective efforts, things tend to go well and to move in the right direction. When individual and social values lose their attractive and cohesive force, things tend to go poorly and disintegration, disorganization, and dissolution set in. Values in both biotic and cultural systems function as "attractors" or "trajectories" of self-organization. When the attraction of a value or value constellation no longer sustains orderly internal self-organization, disorder results; a shift toward chaos occurs, sometimes abruptly. Whether in an organism, a genome, an ecosystem, the brain of a human being, a human organization, or a human-created system (e.g., an economy in which goods and services are freely exchanged, or an "internet" in which information is freely exchanged), capacity for self-organization generates order and tendency toward well-being.

On this account, humans are not the origin of all value in nature. Only humans are conceptual valuers, but this does not entail the conclusion that only humans are valuers and have a well-being. Levels of complexity exist in nature, and so do levels of intelligence and awareness. Jonathan Schull said this about plants: "Gene pools in evolving populations acquire, store, transmit, transform, and use vast amounts of fitness-relative information. … The information-processing capacities of these massively parallel distributed processing systems surpass that of even the most sophisticated man-made systems."[41] Such intelligence has a physiochemical basis, albeit one far removed from the cognitive abilities of a human being equipped with multiple senses and a nervous system controlled by electrochemical reactions occurring in the billions of cells in a human brain. There is a big difference, but the difference is not that one involves "matter" and the other "mind"; the difference, rather, is one of levels of evolved self-organizing capacity and differentiated function. Rudimentary awareness in simple self-organizing organisms has led to progressively sophisticated awareness in progressively complex self-organizing systems. Over several billion years, countless trillions of organisms have embarked on paths of self-actualization in environments in which self-organizing ecosystems have delineated and sustained mutually

---

41 Shull is quoted by Holmes Rolston in "Naturalizing Values: Organisms and Species," in *Environmental Ethics: Readings in Theory and Practice*, Louis P. Pojman, ed. (Wadsworth, 2001), p. 83.

advantageous relationships. Countless species have appeared on the scene, including vertebrate organisms in whose lives value is associated with experiences of pleasure and pain, and humans in whose lives value is associated as well with reasoning and its cultural creations.

With the appearance of *Homo sapiens*, awareness has itself evolved to the point where the story of life can be told. A forthright telling of the story acknowledges that the capacity to value is not unique to human life and that valuation in human life, as advanced as it is and may yet become, is an evolutionary product of valuation in very many nonhuman lives over a very long period. A theory that altogether denies the reality of valuing process in other organisms and in evolving ecosystems is an anthropocentric myopia lacking current scientific credentials. Members of *Homo sapiens* are the inheritors of capacities found throughout the biotic order. A functioning human brain, the seat of rationality, is itself a highly complex self-organizing system many millions of years in the making.

The Value of Nature for Humans

The discussion of value has focused so far on values *in* and *throughout* nature. It now shifts to the value *of* nature—nature's *extrinsic value*—for human need fulfillment. Agriculture, fishing, and forestry harvest resources directly from nature. Plant species domesticated from wild species are the source of most of the calories consumed in diets and a primary source of genetic materials used for the prevention and treatment of disease. A third of all prescription drugs are derived from chemical compounds found in nonhuman nature, and pharmaceutical companies are aggressively "bioprospecting" to find other potentially useful compounds in rainforests and other ecosystems. Biotechnology is yielding an astonishing range of new and useful products, as are efforts to mimic processes in nature (biomimicry) and to miniaturize technologies that replicate processes at the microscopic level (nanotechnology).

In these and other ways, nature provides the wherewithal of human life and is the source of raw materials used to produce the products traded in economies around the world. The value of these inputs are explicitly acknowledged when they are treated as production costs and reflected in market prices. Less acknowledged but more fundamental are the myriad ecological services performed by nonhuman nature, including nutrient cycling, pollination, air and water purification, climate control, drought and flood control, and regulation of atmospheric chemical composition—tasks that nature provides for free. Even if evolving science and technology could one day lead to human substitutes for many of nature's ecological services, the associated cost—conservatively estimated to exceed the total of the world's combined GDPs[42]—would far exceed humanity's ability to pay. As discussed elsewhere, the failure of markets and businesses to account for the depletion and degradation of natural capital is a primary cause of some of our most serious environmental and ecological problems.

Nature as Mentor

Our sciences and technologies are based on what we have learned *about* nature, but much of our wisdom is what we learn *from* nature. Albert Einstein told us to "Look deep into nature, and then you will understand everything better." Leonardo da Vinci remarked that "Human subtlety will never devise an invention more beautiful, more simple, or more direct than does Nature, because in her inventions, nothing is lacking and nothing is superfluous."

Nature has experimented for billions of years with arrangements that enable diverse life to survive, coadapt, and reproduce, and only hubris denies that nature is humankind's most experienced and most reliable mentor. Indigenous peoples understood this truth intuitively, as have Taoists and the followers of numerous other cultural and religious traditions that see humans as inseparably linked to "nature's ways." Michael Pollan saw nature as the gardener's most reliable mentor: "If you think of evolution as a three-and-a-half billion-year-long laboratory experiment, and the gene pool as the store of information accumulated during the course of that experiment, you begin to appreciate that nature has far more extensive knowledge about her operations than we do."[43]

While science is practiced to learn about nature as a means of control (the "Baconian Project"), we are well-advised to look to nature for guidance in organizing human affairs. In her book, *Biomimicry*, Janine Benyus explains why humans would do well to pay heed to nature's ways, which she summarizes in a series of observations: nature runs on sunlight, uses only the energy required to meet basic needs, fits form to function, recycles everything and wastes nothing, rewards cooperation, banks on diversity, demands local expertise, and curbs excesses and unsustainability.[44] Many examples can be

42 Robert Costanza et al., "The Value of the World's Ecosystem Services and Natural Capital," in *Nature* 387 (1997), p. 253.

43 Michael Pollan, *Second Nature: A Gardener's Education*, Op. cit.

44 Janine Benyus, *Biomimicry* (William Morrow, 1997), p. 7.

given of how inventors, designers, and fabricators in numerous fields mimic processes or otherwise incorporate insights derived from nature, including zero-waste and cradle-to grave concepts in product design and manufacturing and such architectural design concepts proposed in McDonough's and Braungart's Hannover principles, Van der Ryn's and Cowan's five principles of ecological design, and John and Nancy Todd's principles of ecological design, among others.

Summing Up

If an attitude of respect toward nature were to take hold in the mind of individuals and organizations throughout society, we might have a "counter-driver" capable of countervailing the seemingly unstoppable forces that have gotten us into the environmental and ecological predicaments we are in, including the widespread trust placed in society's vast technological enterprises and allegiance given to the "technological imperative," the faith that an "invisible hand" operates in unregulated markets to promote the greatest good for the greatest number, the belief ecological services are really "free," and the prevalence of moral and legal systems that deny moral standing to nonhuman life. Such beliefs are widely held in Western society and are spreading throughout the world, and will most likely not be seriously challenged unless and until an attitudinal change begins to take hold in the minds of many. Without groundswells of public sentiment, the initiatives needed to effectively address environmental and ecological problems will very likely remain piecemeal and easily derailed by powerful interests motivated to maintain the status quo.

Absent a paradigm shift in how humans view themselves in relation to the natural world, it is hard to see how catastrophic consequences can be averted. Many people express confidence in the capacity of technology to see us through and optimistically believe that the inventive and creative mind of humans will find technological solutions to whatever problems are looming on the horizon. The problem with this position is that technology is deployed to serve human purposes and, more narrowly, national and corporate interests.

The irony is that there are solutions to many of the ecological and environmental problems we face. Along with other thoughtful observers, Lester Brown believes that "all the problems we face can be dealt with using existing technologies" and "almost everything we need to do to move the world economy onto an environmentally sustainable path has been done in one or more countries."[45] The problem is that the majority of those making the decisions are unwilling or unable to implement at-hand solutions. If worst-case scenarios associated with climate change and biodiversity loss come to pass—such scenarios become more likely with each passing year that individuals and nations fail to take preventive action—solutions unfavorable to humanity will be imposed. An alternative outcome might be effected by a worldwide public outcry insisting that preventive action be taken by decision makers in industry and government at all levels, together with personal commitments to do what is needed at the individual level. Unless such a groundswell were grounded in a paradigm shift in people's attitudinal orientation toward nature, and were fueled by an understanding and fear of what we all stand to lose, needed changes seems unlikely.

As a way of summarizing the previous discussion of factors contributing to an attitude of respect for nature, the following rhetorical questions are posed:

- We attribute value to the human communities that sustain and enrich us. Should we not attribute value to the larger biotic communities that sustain and enrich us?
- We value institutions that facilitate cooperation and mutually advantageous relationships in society. Should we not also value cooperative and mutually advantageous relationships in ecosystems?
- We attribute value to intelligence, beauty, and excellence in human life? Should we not also value the analogues and precursors of these values in nature?
- We value the wisdom of mentors in society. Should we not also value nature's wisdom, nature as mentor?
- We view human history with a sense of awe, wonder, and respect. Should we not also regard nature's much longer and certainly no less storied history with awe, wonder, and respect?

45 Lester Brown, *Plan B 2.0: Rescuing a Planet Under Stress and a Civilization in Trouble* (W. W. Norton, 2006), p. 17.

- We feel a special connection with the local places we come to know intimately and feel "at home." Should we also not feel a connection with the broader natural environments in which these places are embedded, derive their distinct characteristics, and are sustained?

"Yes" answers to these questions parallel previous discussion of contributing elements to an attitude of respect. Affirmative answers constitute an acknowledgement that there much in nature that is worthy of respect.

FOUR ENVIRONMENTAL VIRTUES

As discussed in the Chapter 1 introduction to environmental virtue ethics, numerous traditional virtues are relevant to the discussion of right attitudinal orientation toward nature, including such virtues as benevolence, frugality, humility, gratitude, moderation, discipline, and perseverance. Vices such as arrogance, carelessness, greed cruelty, indifference, pride, wastefulness are no less relevant in this discussion.

Here, though, the emphasis is on four primary virtues: respect, humility, gratitude, and responsibility. In view of the ominous environmental and ecological challenges we face, these four virtues should, it is contended here, be at the center of a much-needed shift in people's attitudinal orientation toward nature.

Respect

Much has already been said about respect. Without being overly repetitive, it has been emphasized that respect for nature derives from numerous kinds of doing, thinking, and experiencing, including reawakening and rekindling feelings of connectedness with the land and its many moods, inhabitants, and voices; experiencing the awe, beauty, and wonder that direct experiences in nature evoke; coming to feel at home not only in the local places we live but the broader natural environment in which familiar places are situated and from which they derive many of their distinctive features; contemplation of life's vast evolutionary journey over some three-and-a-half billion years; setting aside mechanistic and reductionist frames of reference to enable a fuller understanding of capacities in biotic systems to self-organize, adapt, recover, create, and evolve, often in increasing levels of complexity and diversity; appreciating that ancient life forms reside and function inside our bodies, constituting an essential part of what it means to be human; and coming to grips with what we stand to lose if human-caused climate change and massive biodiversity loss go unchecked.

If one virtue stands out as a preeminent environmental virtue, it is respect grounded in the belief that value inheres in nature itself. A world view that associates all value with human value and, more narrowly, economic value cannot engender genuine respect. Acknowledgment that we need nature for survival and the many good things that nature provides is not the same as respecting nature. Again, one does not respect something that one believes has no value in its own right.

Humility

Humility has been heralded by many writers in many contexts as a primary virtue—certainly by Thoreau and Muir, by scientists contemplating the wonders of the cosmos and nature's intricate workings, by religious teachers reflecting on divine transcendence and appropriate religious orientation, by those reflecting on the great achievements and failures of life in its three-billion-year-long evolutionary history, by adventurers and explorers in the mountains and on the seas, and by many less illustrious people growing gardens, caring for crops and animals, and committed to living in harmony with the places they inhabit and call home.

Many cultures extol the virtue of humility. Confucius said that "humility is the solid foundation of all the virtues." Jesus taught the central importance of humility, as have Christians throughout the centuries, including Saint Francis of Assisi, who professed that one cannot but be humbled when one beholds the wonders and beauties of God's creation. "Praise and bless the Lord," Saint Francis said in his *Cantle to the Sun*, "and give thanks and serve him with great humility."

One does not exaggerate one's power or significance when one is immersed in wild nature, when one encounters wonders that one cannot decipher, or when one stands before forces one cannot control. Humility in its most elemental form comes from the realization that nature's powers and achievements transcend human powers and achievements, and that its history transcends human history. Scientists speak of humility of this sort when they contemplate the deepest mysteries of the universe, as do deeply religious people when they contemplate divine transcendence.

A person with humility is a person who rightly assesses one's own powers and potentiality (influence, significance, performance, contributions) in relation to ideals and the powers and potentialities one finds around oneself. Humility comes from honestly sizing oneself up when comparisons of this sort are made, and it is revealed when one addresses and represents oneself to others with appropriate modesty. Humility is not self-aggrandizement, nor is it self-deprecation or insincere humbling; it is, rather, a demeanor grounded in the realization that there are forces, ideals, and time frames in comparison with which one comes up short. In this realization, arrogance and hubris are throttled and humility sets in.

In reflecting on mountains and their implications in his book, *Mountains of the Mind*, Robert writes that "Most of us exist for most of the time in worlds which are humanly arranged, themed, and controlled. One forgets that there ae environments which do not respond to the flick of a switch or the twist of a dial." Mountains, he says, correct this amnesia. "By speaking of greater forces than we can possibly invoke, and by confronting us with greater spans of time than we can possibly envisage, mountains refute our excessive trust in the man-made. They pose profound questions about our durability and the importance of our schemes. They induce, I suppose, modesty in us."[46]

Numerous writers have commented on the folly of thinking that science and technology have given humans the upper hand in life's evolution, and that we are now, or soon will be, in control of our destiny and that of other life. We are reminded that vast forces such as climate change, continental drift, massive volcanic eruptions, and meteor collisions are "game changers" for all life; that dinosaurs and other dominant species have had their moments in the sun but are now long gone; that the more we think we know, the more we realize that there is more to be known; and that it is dangerously arrogant to think that nature can be redesigned and engineered to suit human purposes. We need to live safely and comfortably in our environments, but mastery of nature is not in the cards. Stephen Gould suggests that we treat the earth as we would want it to treat us. "She holds all the cards and has immense power over us—so such a compact, which we desperately need but she does not at

46 Robert MacFarlane, *Mountains of the Mind* (New York: Vintage Books, division of Random House, 2004), pp. 275–76.

her own time scale, would be a blessing for us, and an indulgence for her. We had better sign the papers while she is still willing to make a deal."[47] Thinking that humans have the ability and right to direct the trajectory of life on the planet and to safely harness geophysical forces is a morally flawed illusion propped up by arrogance.

Gratitude

Historically, gratitude, such as humility, has been closely linked with religious traditions. In Judaism, all good things come from God, and giving thanks for God's blessings is an imperative. Psalms 30:12 says, "O lord my God, I will give thanks to you forever," and the Jewish prayer book begins with this simple expression of gratitude: "Thank you, God, for returning my soul to my body." Expressions of gratitude for God's blessings also permeates Christian teachings and formal religious practice. The medieval mystic Meister Eckhart suggested that "If the only prayer you said in your whole life was 'thank you,' that would suffice," and Martin Luther called gratitude "the basic Christian attitude." In Islam, gratitude is expressed to Allah in each of five daily prayers and fasting during the month of Ramadan, and is intended to strengthen the resolve of believers to submit to God and to put believers in a state of gratitude for the abundance that God provides.

Although gratitude may be less commonplace in today's secular world, it has experienced somewhat of a comeback with the rise of the positive psychology movement. Studies indicate that gratitude has one of the strongest links with mental health of any personality trait. In one study, participants were randomly assigned to one of six therapeutic interventions designed to improve perceived overall quality of life. One intervention involved having participants write and deliver a letter of gratitude to someone in their lives. Results showed a rise in happiness scores by 10 percent and a corresponding decrease in depression scores, results that lasted up to one month. Comparable but longer-lasting effects were found to be associated with the keeping of "gratitude journals" in which participants wrote down three

---

47 Stephen J. Gould, "The Golden Rule—A Proper Scale for Our Environmental Crisis," in *Reflections on Nature*, Lori Gruen and Dale Jamieson, eds. (New York: Oxford University Press), p. 39.

things they were grateful for every day.[48] Other studies provide evidence that the expression of gratitude may be a uniquely important contributor to one's sense of well-being.

Aboriginal people everywhere have expressed gratitude for nature's bounty, but relatively few moderns do so today. Homes, cars, gas, food, beverages, phones, computers, and other household and personal items are purchased from stores, and public and private service providers make electricity, water, phone service, radio, television, and the internet readily available. In today's world, there are a vast array of intermediaries between us and food on the table, or between us and light emanating from a light bulb. City dwellers are remotely aware that life's necessities come from nature, but for the most part do not care until breakdowns occur in production and distribution systems and immediate needs are frustrated, or until health and safety problems crop up, or the parks, beaches, or other natural places where they recreate disappear or deteriorate. If gratitude is perfunctorily expressed on a day of thanksgiving or voiced at family mealtimes, it is generally directed to a divine provider, where divinity is conceived as agency and nature is conceived as an instrument of God's providence.

There are exceptions to this picture, of course. Urban gardeners appreciate the land and its soils, plants, and pollinators and recognize that gardening success is owed as much to the vitality of their plants as the nurture they provide. Also, many people appreciate the flora and fauna of urban parks and nature preserves. People who go outside cities for sightseeing and recreation appreciate that there are relatively unspoiled natural places, even if, as visitors, they may take these places for granted.

The remoteness of nature in the daily lives of urban dwellers is part of the explanation why heartfelt expressions of gratitude for nature's contributions to our lives are hard to find. Another is the all-too-human tendency to simply take things for granted. As Aldous Huxley put it: "Most human beings have an almost infinite capacity for taking things for granted." But still another part—a critical part—is that nonhuman things in nature are viewed by many people as not the kinds of things toward which gratitude is appropriately directed. If nature is thought to have value only insofar as and to the extent that it has value for humans and is otherwise devoid of value, there is nothing to which gratitude is obviously attached. One does not show appreciation

48 M. Seligman, T. Steen, A. Park, and C. Peterson, "Positive Psychology Progress: Empirical Validation of Interventions," in *American Psychologist*, 60, 410–421.

for something one thinks has no well-being, purpose, intelligence, striving, achievement, or value of its own; one may have a dim realization that good and necessary things come from nature, but there is little, if any, inclination to be grateful for the nonhuman source and to see it as a gift.

We have good reasons for being grateful for nature's contributions to our lives—not only for the manufactured goods that make their way to us as food and amenities to make our lives more safe, secure, and comfortable, but also for the many venues it provides for recreational and aesthetic activities, its value as mentor, and its provision of essential ecological services. There are many things in the natural world to which gratitude meaningfully attaches. If it is meaningful to be grateful for what organizations contribute to our lives, or for divine providence, it is meaningful to be grateful for life-sustaining capacities in nature *writ large* and for the tangible benefits that nature in all of its particularity provides.

Responsibility

A variety of meanings are associated with the concept of responsibility. In one sense, the term "responsibility" is used as a synonym for "obligation": to have a responsibility is to have an obligation. In a second sense, "responsibility" is used to convey that people are accountable for their actions, and that penalties for irresponsible acts are justifiable. Here the terms "accountable" and "responsible" are used interchangeably. A third meaning refers to "taking responsibility" and "holding accountable," where the former means accepting that that one is responsible (accountable) for one's own actions and the latter means holding others responsible (accountable) for their actions. Putting these meanings together, a responsible person is a person who (1) acts in accordance with one's responsibilities (obligations), (2) takes or accepts responsibility for one's own actions, and (3) holds others accountable for their actions.

Laws and law enforcement hold people accountable for practices that deplete and degrade natural resources (primarily to protect human health and safety and ensure a sound economy) and for practices involving the mistreatment of animals (particularly vertebrate animals). But laws and law enforcement are not enough. No less needed and important is the exercise of individual responsibility: of people—many people—not acting in ways that are clearly detrimental to the environment and animal welfare; not shirking or avoiding personal responsibility or looking for excuses; taking responsibility for finding out what one can do to make a positive difference; holding others responsible, including decision makers in government and industry; joining and supporting activist organizations working to reduce waste, conserve wildlife and biological diversity, protect land and water resources, and promote animal welfare; and making lifestyle changes to reduce one's own ecological footprint. There is no dearth of information about what one can and should do personally to help reduce waste, promote biodiversity, and do one's own part in helping to prevent the dangerous consequences of continued global warming. What is most needed, and oftentimes most lacking, is the exercise of personal responsibility.

5 ANTHROPOCENTRIC PERSPECTIVES

This chapter critically examines arguments for and against three central anthropocentric claims: that all and only humans have moral standing;[1] that value originates in human experience and nowhere else; and that a fully adequate environmental and animal ethic can be developed from an "enlightened" anthropocentric theoretical framework.

ANTHROPOCENTRIC THESES

Only Humans Have Moral Standing

Virtually all moral theorists today begin with the premise that humans have moral standing: of course, humans are morally considerable in their own right, and of course the harm that humans inflict on other humans raises the question of the moral permissibility of that harm. Some societies and theorists have contended that only *some* humans count morally on the grounds that they are inherently superior to other humans, but such claims have been roundly denounced as morally unacceptable prejudice, and social movements have been mobilized and laws enacted to eliminate this prejudice. The question to be debated in the following is not whether *all* humans should be regarded as having moral standing, but whether *only* humans should be regarded as having moral standing,

Religious arguments for the view that humans have moral standing are rooted in the belief that God created a hierarchical world populated by

1 The meaning attached to the concept of moral standing employed in this work is discussed in Chapter 1.

beings whose level of worth and being are measured by how close they are to God's perfection.[2] Only humans were thought to be "created in the image off God": only humans have a soul (possess spiritual substance) and are endowed with the powers of reason and moral agency. Stripped of religious explanation, the ability to reason (rationality) was singled out as the core meaning of what "being created in the image of God" means. While rationality is variously defined, its core meaning refers to the ability of humans to entertain beliefs and support them with reasons, including beliefs about what is true and false, right and wrong, and good and bad, together with the ability of humans to accept responsibility for doing right and the reasonableness, thus, of others holding humans accountable for doing right. This, essentially, is the position developed by Plato and Aristotle, reiterated by Kant, and held by many contemporary anthropocentric theorists. Rational capability is heralded as eminently worthy of respect, although humans fall short of their realizing their potentialities as rational beings.

For an anthropocentrist, then, the ability to reason is taken to be the essence what it means to be human and is regarded as a value intrinsic to human nature. Ability to reason merits moral respect, and is the only acceptable criterion of moral standing.

Only Humans Value and Have Inherent Value

Anthropocentric theorists contend that humans and only humans value things—because only humans have the ability to entertain value concepts and make value judgments. If there is no conceptual valuer, there is no value. Value originates in human experience, and nowhere else.

Accordingly, the anthropocentrist also contends that a nonhuman entity (an animal, a plant, a genomic kind, an ecosystem) has value only if it judged to have value by a human, where the "only if" implies that nonhuman entities do not have inherent (intrinsic) value in their own right. Apart from their instrumental or aesthetic value for humans,[3] nonhuman entities are valueless; a world without humans would be a world without value.

2 In the Middle Ages this view was articulated in various versions as the Great Chain of Being, with key elements drawn from the metaphysics and ethics of Plato and Aristotle.

3 While an anthropocentric value theorist denies that biotic entities have human-independent intrinsic (inherent) value, it is often claimed that objects that occasion positive aesthetic

Implications of "Enlightened Anthropocentrism" for Environmental Ethics

Anthropocentric theorists readily acknowledge that we have prima facie obligations not to mistreat animals, not to pollute the land, air, and water, not to destroy habitats and drive endangered species to extinction, not to hasten the pace of global warming, not to consume renewable resources in excess of replenishment rates, and, in general, not to engage in harmful environmental and ecological practices. For the anthropocentrist, however, these obligations derive from considerations of what is necessary for human survival and well-being, not from considerations predicated on the notion that value resides in nature independently of human life and culture. Disagreement arises over specific obligations in specific cases, but there is no disagreement among anthropocentrists that environmental obligations (indeed, all obligations) derive from considerations of human well-being and are ultimately obligations owed to humans, not to nonhuman biotic entities. If some obligations imply correlative rights, the rights are human rights, not the rights of nonhumans.

William Godfrey-Smith neatly summarized four lines of argument for exercising good environmental stewardship: the *cathedral argument,* highlighting the opportunities that wilderness areas afford for spiritual renewal and aesthetic delight; the *laboratory argument,* reminding us that the opportunity to study nature in all its diversity enables us not only to discover biological laws, the transgression of which puts human life in peril, but also to derive the compounds needed to make pharmaceuticals that prevent, treat, and cure human disease; the *silo argument,* reminding us that good things come from nature, including food, water, energy, and backup stockpiles of genetic diversity in case something goes suddenly wrong with monoculture agriculture; and the *gymnasium argument,* focusing on the many athletic and recreational opportunities that nature provides.[4] The common denominator in all these reasons is that they are reasons for securing and safeguarding *human* well-being. Richard Watson put it simply and bluntly: "There is very good reason for thinking ecologically, and for encouraging human beings to

---

experiences—for example, the experience of beauty, awe, and wonder—have intrinsic value, whether or not they also have instrumental or use value for humans.

4 Geoffrey-Smith himself thought that human-centered reasons for protecting the environment, while compelling, were not enough; also needed, he thought, were reasons rooted in the recognition of intrinsic or inherent values in nature.

act in such a way as to preserve a rich and balanced planetary ecology: human survival depends on it."[5]

Bryan Norton, among others, makes the case for an "enlightened" anthropocentrism: an anthropocentrism that takes long-term perspectives and does not overlook the needs of future generations, that does not reduce all human value to economic value, that does not look for a single moral principle to justify and unify pluralistic principles, and that does not seek to justify its human-centered orientation on the claim that humans are inherently superior to other life. He argues that an anthropocentrism, thus oriented, is fully consistent with sound environmental and ecological practices. In this connection, he advances what he calls the "convergence hypothesis": that anthropocentrism and non-anthropocentric thinking, if informed and fully worked out, would yield the same resource management policies and practices. In his words: "At least in the important area of moral obligations regarding fauna and flora, the policy implications of a broadly formulated and farsighted anthropocentrism and non-anthropocentrism will be identical."[6]

Norton was a participant in the heated exchanges among animal rights advocates, environmentalists, and anthropocentrists during the latter half of the twentieth century and sought to develop a practically useful framework that avoided counterproductive vitriol. He summarized his position as follows:

> Rather than *reducing* pluralistic principles by relating them to an underlying value theory that recognizes only economic preferences or 'inherent' value as the ontological stuff that unifies all moral judgments, I have sought integration of multiple values on three scales of human concern and valuation [at the scale of the individual, the community, and the global], choosing pluralism over monism and attempting to integrate values within an ecologically informed model of human habitat. ... We humans will understand our moral responsibilities only if we understand the consequences of our actions as they unfold on multiple scales, and the human community will only survive to further evolve and adapt if we learn to achieve individual welfare and justice in the present in ways that are less

5 Richard Watson, "A Critique of Anti-Anthropocentric Ethics," in *Environmental Ethics: Readings in Theory and Application*, Louis P. Pojman, ed. (Wadsworth, 2000), p. 167.

6 Bryan G. Norton, "Environmental Ethics and Weak Anthropocentrism," in *Environmental Ethics*, 6, pp. 131–148.

disruptive of the processes, evolving on larger spacio-temporal scales, essential to human and ecological communities. ... I endorse the broadly Darwinian approach both to epistemology and morals proposed by the American pragmatists, together with its rejection of the view that value exists independently of humans in nature ("a realist moral ontology") and its reduction of pluralistic principles to a single principle and all human value to economic value.

CRITIQUE OF ANTHROPOCENTRISM

Opposition to Claims of Human Superiority

Many of the criticisms of anthropocentrism have been directed at the version that asserts human *superiority.* Claims of superiority have been associated with the belief that only humans are created in the image of God, or that only humans have a mind, or that humans are the crowning achievement of an evolving natural order. If the explanation is that only humans have a mind, a recurring objection is that it presupposes a mind-body dualism that has been widely rejected in contemporary science and philosophy. Various writers have observed that the mind-body dualism was not a Judeo-Christian concept until Platonic metaphysics was incorporated into Neoplatonic theologies and Aristotelian metaphysics was incorporated into the theology of Thomas Aquinas, only later to regain philosophical status in the writings of Descartes. As previously noted, the rationality criterion is generally taken by most anthropocentrists today as the only acceptable criterion of moral standing, and philosophical discussions in this context are generally decoupled from claims of superiority rooted in mind-body dualism or theological explanations.

Critics have also been quick to point to the claim that humans are superior simply because they belong to the species *Homo sapiens*—that such a claim is no less baseless than the claim that men are superior to women because they belong to the class of males, or whites are superior to nonwhites because they belong to the Caucasian race. Class membership does not establish superiority.

But the discrimination does not have to be this blatant. Authors in the feminist tradition are critical not only of the male/female dichotomy but other dualisms as well—mind/body, reason/emotion, objective/subjective, human/nonhuman, culture/nature—where the first-mentioned characteristics in each of these pairs is characterized as "masculine" and superior to each of their complements, which are characterized as "feminine" and inferior. All feminist writers oppose spurious claims of superiority and inferiority to justify the domination of women by men, and ecofeminists oppose the domination of nature by humans. Karen Warren put it this way: "Ecofeminists insist that the sort of logic of domination used to justify the domination of humans by gender, racial, ethnic, or class status is also used to justify the domination of nature."[7] Ecofeminists contend that species discrimination is as deeply embedded in society as gender and racial discrimination, and their aim is to expose and weed it out.

Problems with the Rationality Criterion

Another frequently made criticism of anthropocentrism is that its defenders cannot come up with a version of the rationality criterion that exludes all nonhumans but includes all humans. As Peter Singer, Tom Regan, and other vertebracentric theorists have argued, a criterion of moral standing that excludes sentient animals also excludes infants, elderly persons with dementia or Alzheimer's disease, imbeciles, and persons who have suffered brain injuries. Exclusion of these members of the human population is morally objectionable to all ethicists. But any adjustment of the rationality criterion to extend moral standing to the very old, the very young, and the mentally disabled has the effect of extending moral standing to at least some "higher animals" as well. The mother chimpanzee's ability to learn a sign language and teach it to her baby chimp, an elephant's mourning of the death of a fellow elephant and treating its remains with respect, and the ability of spiders to construct elaborate insect-trapping webs are examples of intellectual, moral, and technical capabilities that exceed those of the very young, very old, and brain-impaired humans. Various writers attempt to get around this objection by saying that all humans have rational potential, but humans with Alzheimer's disease or who are severely mentally disabled or

7 Karen J. Warren, "The Power and Promise of Ecological Feminism," in *Environmental Ethics: Readings in Theory and Application*, ed. Louis Pojman (Belmont, CA Wadsworth, 2001), p. 192.

have suffered irreversible brain damage do not have even the potentiality of rational thought and action.[8]

Valuing and Value Not Unique to Human Life

Biocentric and ecocentric theorists make the case that nonhuman life and life systems have naturally occurring goods of their own, and that it is not meaningless to speak of processes in nature as involving value expression (valuation). Rationality is good for humans, but so is photosynthesis for a plant. Cellular-level genomic expression is good for all organisms. Coadapted fitness and relations of mutual benefit constitute the good of ecosystems and are good for the vast number of individuals and species that inhabit them. Sentient animals care about their lives, have concerns, and exhibit purposeful behavior. Such caring and goal-oriented behavior in sentient animals are characterized as forms of value expression. While such valuation in sentient animals is not conceptual and ideological, it is also not dismissible as crude anthropomorphizing.

It takes human cognition to recognize such goods and naturally occurring valuation, but, as Holmes Rolston says, the recognition does not create the value; it is already there, awaiting human recognition and articulation.

Asserting that humans are superior because they possess a characteristic that nonhumans lack is a begging of the question. As Paul Taylor notes, "To use standards based on human value is already to commit oneself to holding that humans are superior to non-humans, which is the point in question."[9] James Sterba makes the same point.[10] Humans are uniquely able to formulate and solve complex mathematical equations, but they lack the homing ability of a pigeon, the speed of a cheetah, the scent of a wolf, the visual acuity of an

8 Carl Cohen replies that brain-damaged humans are appropriately regarded as morally considerable even though they lack the capacity to make moral judgments. His argument is that the test for moral judgment "is not a test to be administered to humans one by one, but applied to the capacity of members of the species in general." See Carl Cohen, "The case for the use of Animals in Biomedical Research," in *New England Journal of Medicine*, vol. 315, issue 14, October 1986, pp. 865–870.

9 Paul Taylor, "Biocentric Egalitarianism," in *Environmental Ethics: Readings in Theory and Application*, ed. Louis Pojman (Wadsworth: Belmont, CA, 2001), p. 108. Taylor's biocentric ethical theory is examined in some detail in the next chapter.

10 James Sterba, "Environmental Justice: Reconciling Anthropocentric and Non-anthropocentric Environmental Ethics," in *Environmental Ethics: Readings in Theory and Application*, ed. Louis Pojman (Belmont, CA: Wadsworth, 2001), p. 179.

eagle, and the photosynthetic power of a plant. Mathematical problem solving is good for humans, but not for cheetahs and pigeons. Photosynthesis is good for plants, but not for humans and wolves. To take what is good for or about humans as a basis for asserting human superiority and exclusive moral standing begs the question. The argument presupposes the conclusion that it purports to establish.

Violation of the Principle of Equal Consideration and "Speciesism"

Peter Singer's concern with anthropocentrism is that it violates a basic principle of justice: the principle of equal consideration. The argument goes like this: (1) Humans and other sentient animals similarly experience pain and have a like interest in not being in pain and the emotional distress that pain causes; (2) Like interests ought to be given like consideration in moral deliberation (the principle of equal consideration); and (3) not taking the pain and suffering of animals into moral account and not giving equal weight to similarly experienced human and nonhuman pain and suffering violates the principle of equal consideration. The anthropocentric denial of moral standing for vertebrates derails the possibility of taking the like pain and suffering of sentient animals into moral consideration, and is therefore unjust and discriminatory.

In Singer's words: "If a being suffers, there can be no basis for refusing to take that suffering into consideration. ... The principle of equality requires that its suffering be counted equally with like suffering of any other being." Singer calls the failure to take an animal's pain and suffering into moral consideration "speciesism" and argues that it is no less discriminatory than racism and sexism.[11]

The charge of species discrimination is particularly unpalatable to the humanist who strenuously opposes any criterion that limits the scope of moral considerability to a subset of human beings. Humanists insist that moral standing should not be based on gender, race, nationality, or any other criterion that gives special moral status to *some* human beings. These beliefs have inspired activist movements that have liberated slaves, women,

11 Singer's utilitarianism does not require that animals are given equal treatment, only that their pain and suffering be taken into consideration and given like consideration in moral deliberation about what constitutes right treatment.

and other historically marginalized groups. In view of these achievements, there is a startling irony in the charge that anthropocentrism is another form of discrimination: that the "speciesism" implicit in a claim of special moral status for human beings is no more justified than sexism, racism, or any other "ism" that claims special status for a subgroup of humans. The humanist insistence that *all* humans have moral standing has been a boon to humankind. The claim that *only* humans have moral standing has opened a floodgate of criticism from animal welfare theorists and activists.

Long-Term Environmental and Ecological Decline

As noted earlier, Norton argues that, from both a theoretical and a practical standpoint, non-anthropocentric perspectives are unnecessary: that an "enlightened" anthropocentrism is all that is needed for adequate environmental and ecological protections. Critics question this assertion and the optimism it embodies; they are concerned that anthropocentrism's denial of nonhuman moral standing, together with its denial that value inheres in nature itself, will lead to continuing environmental and ecological decline to the detriment of all planetary life. The thrust of this line of criticism is that an anthropocentric mind-set is incapable of sustaining the kind of "enlightened" humanism that Norton calls for.

One line of criticism is that an environmental ethic requires a deeper sense of respect than vertebracentrism can provide. If nature is thought to be valueless apart from its value for humans, there is no basis for *respecting* nature; nothing to which an attitude of respect meaningfully attaches. One does not truly respect something one believes has no intrinsic worth, and one's inclination to act respectfully is blunted. While Norton calls for a pluralistic value theory that recognizes the full extent of nature's value for humanity and rejects evaluation methodologies that reduce all human value to economic value, it is questionable whether the anthropocentric perspective alone, as "enlightened" as it may be, is enough to stem the seemingly unsustainable practices in society that contribute to biodiversity loss, global warming, and the host of other environmental and ecological problems we face.

Katie McShane articulates another line of argument for questioning whether an anthropocentric outlook is capable of curbing continuing environmental and ecological decline. In her words: "I argue that even if we grant the truth of Norton's convergence hypothesis, there is still good reason to worry about anthropocentric ethics. Ethics legitimately raises questions about how to feel, not just about which actions to take or which policies to adopt. From the point of view of norms for feeling, anthropocentrism has very different practical implications from non-anthropocentrism; it undermines some of the common attitudes—love, respect, awe—that people think it appropriate to take toward the natural world."[12]

When policy deliberations start with the premise that only humans count morally, or that value does not exist apart from human life and culture, harm to nonhuman life and life systems is easily overlooked, trivialized, or dismissed. When a development proposal competes with a preservationist alternative, for example, the preservationist alternative prevails only when the net human benefits of not developing are greater than the net human benefits of developing—a judgment that rarely occurs in practice, particularly when associated benefits and harms are expressed in monetary terms, the planning horizon does not extend to the next generation and the one after that, and future costs and benefits are discounted.

Numerous writers have pointed out the inadequacies of cost–benefit analysis when monetary values are assigned to such values as human health and safety, let alone such things as biodiversity and wilderness tracts. In the absence of direct market values for these kinds of things, market proxies are devised. The value of biodiversity becomes, for example, what people would be willing to pay to avoid its further loss. But however monetary values are derived, the underlying problem is the anthropocentric contention that only the interests and well-being of people matter in deliberation, not the well-being of animals, plants, species, and biotic communities.

Notwithstanding the problems associated with the anthropocentric perspective in environmental ethics, it must be acknowledged that human-centered reasons for protecting the environment are compelling reasons, and that these reasons have carried the day on many notable occasions. Important environmental protection and conservation laws have been adopted because international bodies, governments, nongovernmental organizations, and ordinary citizens have agreed that considerations of human survival well-being require these laws. In the United States there are

12 Katie McShane, "Anthropocentrism vs. Non-anthropocentrism: Why Should We Care," in *Environmental Values* 16(2), pp. 169–85.

the 1964 Wilderness Act, the 1969 National Environmental Policy Act, the 1970 Clean Air Act, the 1972 Clean Water Act, and the 1973 Endangered Species Act. At the international level, a notable achievement was the United Nations' 1985 Vienna Convention on the Protection of the Ozone Layer and its follow-up 1987 Montreal Protocol, ratified by many nations (including the US) and major corporate interests (including DuPont). Another major achievement was the Paris Agreement on Climate Change agreed to in December 2015. But there have also been notable setbacks. In the US, the aforementioned environmental legislation has been repeatedly amended in many cases (not all) to relax requirements at the behest of powerful business lobbies, and the Paris Agreement is viewed in some quarters as flawed and floundering. Meanwhile, many (not all) indicators of environmental and ecological health continue to decline.

6 VERTEBRACENTRIC PERSPECTIVES

Vertebrate animals are sentient beings: they variously have the capacity to see, hear, smell, taste, and touch objects in their environments and to feel pleasure and pain in response to visual, auditory, olfactory, and tactile stimuli. Many vertebrates also have at least the rudiments of such mental states as emotion, memory, belief, desire, intentional action, a sense of the future, and self-awareness.

In view of these abilities and characteristics, serious and oftentimes passionate arguments are made by moral philosophers and animal welfare/rights activists for the recognition of the moral standing of sentient animals and for sweeping change in how these animals are treated in meat-producing "factory farms," in research laboratories, in numerous areas of activity in which animals are used for sport and amusement, and in what we eat and wear.

This chapter focuses on arguments for recognition of the moral standing of some or all vertebrates, together with substantive ethical frameworks that have been developed for articulating and clarifying obligations that humans owe to other sentient beings. Emphasis is given to the utilitarian approach of Peter Singer, the Kantian rights-based approach of Tom Regan, the care-based tradition in ethics that emphasizes that feelings of sympathy and compassion are what brings animals under the umbrella of ethics, and David DeGrazia's systematic effort to derive specific practical guidelines for the ethical treatment of animals, based on an examination of the mental life of animals and application of the general principles of nonmaleficence and equal consideration.

The chapter concludes with a brief summary of the animal welfare/rights movement.

THE ANIMAL ETHICS OF PETER SINGER

Pain and Suffering Matter Morally

Singer's argument for the moral standing of sentient animals rests on a series of premises: that pain and the suffering pain causes are intrinsically bad[1] (a value statement), that the most fundamental interest a sentient being has is not being subjected to pain and suffering, that the experience of pain and pain-related suffering are alike for humans and other sentient animals, and that like interests ought to be given like consideration in moral deliberation (the principle of equal consideration). "If a being suffers," Singer states, "there can be no moral justification for refusing to take that suffering into consideration. ... The principle of equality requires that its suffering be counted equally with the like suffering of any other being.[2]

The contention that pain and suffering are alike for humans and other sentient beings is central to Singer's argument and is defended by him at some length. We know our own pain and "nearly all the external signs that lead us to infer pain in other humans can be seen in other species, [including] writhing, facial contortions, moaning, yelping or other forms of calling, attempts to avoid the source of the pain, appearance of fear at the prospect of its repetition, and so on." Singer argues that linguistic ability is not a decisive factor in the determination of whether only humans suffer: "Human infants and young children are unable to use language. Are we to deny that a year-old child can suffer? If not, language cannot be crucial." He continues: "It is surely unreasonable to suppose that nervous systems that are virtually identical physiologically, have a common origin and a common evolutionary function, and result in similar forms of behavior in similar circumstances should actually operate in an entirely different manner on the level of subjective feelings." And "although human beings have a more developed cerebral cortex than other animals, this part of the brain is concerned with thinking functions rather than with basic impulses, emotions, and feeling ... located in the diencephalon, which is well developed in many other species of animals,

1 [3] A hedonistic value theorist holds that the experience of pleasure is intrinsically good (good in and of itself, good for its own sake), and that the experience of pain is intrinsically bad (a disvalue). Although a hedonist regards pain to be intrinsically bad, it is sometimes viewed from an evolutionary and learning standpoint as instrumentally (extrinsically) good.

2 [2] This quotation and those that follow are taken from the second (1990) edition of Singer's *Animal Liberation,* published by Avon Books.

especially mammals and birds." "So … there are no good reasons, scientific or philosophical, for denying that animals feel pain. If we do not doubt that others feel pain, we should not doubt that other animals do so too. … Animals can feel pain."

Humans and other vertebrates have a common interest in not being subjected to pain and suffering. Stones do not have interests, Singer says, but sentient beings do. In his words, "It would be nonsense to say that it was not in the interests of a stone not to be kicked along the road. … A stone has no interests because it cannot suffer. Unlike a stone, "A mouse has an interest in not being kicked along the road because it will suffer if it is." Thus, "If a being suffers, there can be no basis for refusing to take that suffering into consideration. … If a being is not capable of suffering, or of experiencing enjoyment or happiness, there is nothing to be taken into account." To have any interests at all, one must be capable of experiencing pleasure and pain.

Given the primacy of the interest of all sentient animals not to be subjected to pain and suffering, Singer argues that the capacity to experience pain-induced suffering is an appropriate criterion of moral standing and that the anthropocentric criterion is arbitrary or, worse, discriminatory. "Why not choose some other characteristic like skin color?" Singer's answer is that doing so is plainly and unequivocally morally discriminatory. "Racists violate the principle of equality by giving greater weight to the interests of members of their own race when there is a clash between their interests and the interests of those of another race. Sexists violate the principle of equality by favoring the interests of their own sex. Similarly, speciesists allow the interests of their own species to override the greater interests of members of all other species. The pattern is identical in each case."

Singer's Utilitarian Conclusions

Singer's ethics incorporates two basic principles: the principle of utility and the principle of equal consideration. The principle of utility states that one ought to choose that course of action (among alternatives) that is expected to promote the greatest good for the greatest number of affected parties. The principle of equal consideration states that like goods in the lives of those affected ought to be given like consideration in moral deliberation. Substituting the terms "interests" for "goods," Singer states the principle of utility thusly: "One ought to choose and act in a manner that maximizes the interests of all affected parties," in which, he adds, "the interests of every being affected by an action are to be taken into account and given the same weight as the like interests of any other being."[3] In the second edition of *Practical Ethics,* Singer adopts a "journey model" of life, which, he argues, enables one to think of right and wrong actions in relation to the degree that they serve/frustrate the goals of a person's journey in life, and to make needed distinctions between genuine needs and trivial wants and desires.

Singer's utilitarian predecessor, Jeremy Bentham, affirmed that animals other than humans experience pleasure and pain. For Bentham, an action expected to cause animal pain and suffering ought to be taken into consideration in a utilitarian deliberation concerning the moral rightness of the action in question. Hence his famous dictum: "The question is not, Can they *reason?* Nor, Can they *talk?* But, Can they *suffer?*" Singer, as Bentham before him, characterizes pain as intrinsically bad and regards pain avoidance as the most basic interest one can have.

Singer has applied the principle of utility, together with the principle of equal consideration that it incorporates, to argue what he thinks is morally right and wrong on a wide variety of issues, including animal treatment, abortion, euthanasia, infanticide, and world poverty. In regard to abortion, he argued that it is not wrong to kill a human fetus, at least up to at least eighteen weeks, because it has no capacity to suffer or feel satisfaction, has no interests and preferences, cannot be said to have embarked on a life journey, and lacks the essential characteristics of personhood. Singer's conclusion is that there is nothing to weigh against a woman's choice to have an abortion.

In regard to sexual activities between humans and animals, he wrote that "sex with animals does not always involve cruelty" and that mutually satisfying activities of a sexual nature may sometimes occur between humans and animals, in which circumstances they are morally permissible. Tom Regan took strong exception to this position, arguing that the same form of argument could be used to justify having sex with children, and that Singer's conclusion is a consequence of his adopting a utilitarian approach to animal treatment rather than a strictly rights-based one.

---

3 Peter Singer, "All Animals are Equal," in *Earth Ethics,* James P. Sterba, ed. (Upper Saddle River, NJ, Prentice Hall, 2000), p. 53. This article was originally published in *Philosophical Exchange* I (1974), pp. 103–116.

In regard to animal treatment, Singer's general conclusion is that the principle of utility enjoins us "to make radical changes in our diet, the farming methods we use, experimental procedures in many fields of science, our approach to wildlife and to hunting, trapping and the wearing of furs, and areas of entertainment like circuses, rodeos, and zoos."[4] "As a result," he says, "a vast amount of suffering would be avoided."

In making his case against meat-eating and factory-farming practices, Singer argues that horrendous pain and suffering are inflicted indiscriminately and unnecessarily on hundreds of millions of sentient animals and that these practices pose threats to human health, are economically unjustified, are damaging to the environment, and would, if curtailed, contribute significantly to the reduction of Third World poverty. He calls himself a vegetarian and a "flexible vegan": "I don't go to the supermarket and buy non-vegan stuff for myself. But when I'm traveling or going to other people's places I will be quite happy to eat vegetarian rather than vegan."[5] He does not make a blanket condemnation of meat eating, however, arguing that it is permissible on utilitarian grounds to eat farm animals that are reared and killed painlessly and without suffering.

Never reticent about expressing his views publicly, Singer engaged in a debate with Judge Richard Posner of the US Court of Appeals for the Seventh Circuit on the issue of "animal liberation."[6] Posner said his moral intuition tells him "that human beings prefer their own." Further: "If a dog threatens a human infant, even if it requires causing more pain to the dog to stop it than the dog would have caused to the infant, then we favour the child. It would be monstrous to spare the dog." Singer challenged Posner's intuition by arguing that discriminatory practices based on gender, race, ethnic background, and sexual orientation have been justified using appeals to intuition. Posner replied by saying that, in a legal context, facts-driven arguments prevail, not ethical arguments that run contrary to what most people believe. In regard to antidiscrimination law, Posner adds, "facts mounted supporting the case that [there are] no morally significant differences between humans based on race, sex, or sexual orientation." Posner called his approach "soft utilitarianism." In his words, "The 'soft' utilitarianism position on animal rights is a moral

4 Singer, "All Animals are Equal." Op. cit., p. 60

5 Quoted in Dave Gilson's "Chew the Right Thing" in *Mother Jones*, May 3, 2006.

6 Posner–Singer Debate, *Slate*, June 12, 2001.

intuition of many, probably most, Americans. We realize that animals feel pain and think that to inflict pain without a reason is bad." But, he adds, "Nothing of practical value is added by dressing up this intuition in the language of philosophy; much is lost when intuition is made a stage in logical argument. When kindness toward animals is levered into a duty of weighting the pains of animals and people equally, bizarre vistas of social engineering are opened up."

Brian Luke and many ecofeminists believe that Singer's arguments overlook the central role of feelings of sympathy and compassion in moral decision making; Baird Callicott notes that Singer "heaps scorn on the 'romantic sentiment' of sympathy"; and English philosopher Bernard Williams has commented that Singer is "too damn logical," with the consequence that he "leaves out an entire dimension of value." It is perhaps not surprising that Singer's utility- and justice-based conclusions have sparked public outrage in some quarters, including revoked invitations to speak, heckling from protesters before and during presentations, and cancelled donations to universities where he has taught by alumni and other donors.

THE ANIMAL ETHICS OF TOM REGAN

Being a Subject-of-a-Life Matters Morally

Tom Regan proposes a more restrictive criterion of moral standing: the capacity to be an experiencing "subject of a life." Like all vertebrates, mammals feel pleasure and pain, but, more than this, they remember, anticipate events, form social bonds, have a sense of self, and have complex emotions. Their mental lives do not include the full range of cognitive elements found in human experience, but they are subjects of a life concerned about their lives, and there can be little doubt that they suffer in ways comparable to humans.[7] Regan's criterion confers moral standing to all mammals, but not all vertebrates as does Singer's criterion.

Echoing Singer, Regan criticizes the anthropocentric position on moral standing on the grounds that it violates the principle of equal consideration. Mammals and humans are alike in the respect that they are subjects of a life; as us, they have subjective experience with desires, fears, memories, and a

7 See Regan's *The Case for Animal Rights* (University of California Press: Berkeley, 1983) for his description of characteristics attributable to an experiencing subject-of-a-life.

sense of self through time. Because mammals have a complex psychological life, feel pain, and suffer in much the same way we do, and because morally comparable beings ought to be given comparable moral consideration, denial of moral standing to nonhuman mammals is morally unacceptable.

Even more baseless, Regan believes, is the claim that humans are morally considerable because they belong to the species *Homo sapiens*. In *Animal Rights, Human Wrongs*,[8] Regan says this claim is "speciesism in its most blatant form." It is also rationally unacceptable, Regan believes, to base exclusive human moral standing on the controversial claim that only humans have an immortal soul. Although Regan says he hopes that his psychic self continues to exist beyond physical death, he states that it is rationally preferable to resolve moral issues without making more controversial assumptions than are needed." It is also rationally unacceptable, he contends, to claim that humans have moral standing because other animals lack rational capacity, autonomy, or intellect. He argues that an appeal to rationality is no more acceptable than saying that a mature adult has greater moral value than a mentally disabled child because only the former is able to reason and make autonomous judgments.

Regan's Deontological Critique of Singer

Regan insists that humans, indeed all experiencing subjects of a life, have inherent value and therefore are equally deserving of not being treated merely as a means to another person's ends. "Utilitarianism has no room for the equal moral rights of different individuals," he says, "because it has no room for their equal inherent value or worth. What has value for the utilitarian is the satisfaction of an individual's interests, not the individual whose interests they are."[9] For Regan, but not Singer, animals deserving equal consideration also deserve equal treatment.

Mammals are experiencing subjects of a life and, as such, are worthy of moral respect: the kind of respect that not only acknowledges they are morally considerable in their own right, but entitled not to be treated as means to human ends. Regan insists on the kind of treatment for mammals that Immanuel Kant reserved for humans: subjects of a life ought to be treated as ends in themselves and never as a means to human ends only.

8 Tom Regan, *Animal Rights, Human Wrongs* (Rowman & Littlefield, 2003), p. 63–74.
9 Tom Regan, *Animal Rights, Human Wrongs*, Op. cit., pp. 63–64, 89.

Regan straightforwardly declares that he is "an advocate of animal rights—as part of the animal rights movement ... committed to a number of goals, including the total abolition of the use of animals in science, the total dissolution of commercial animal agriculture, [and] the total elimination of commercial and sport hunting and trapping." His animal ethic is absolutist, as was Kant's. He refuses to concede, as Singer does, that traditional animal agriculture is morally permissible if animals are raised and killed painlessly, or that animal testing in laboratories is permissible if it undertaken with the objective of finding a cure for human disease and the pain and suffering associated with testing are minimized. "You don't change unjust institutions by tidying them up," Regan insists. "What's wrong—what's fundamentally wrong—with the ways animals are treated isn't the details that vary from case to case. ... The fundamental wrong is the system that allows us to view animals as our resources, here for us—to be eaten, or surgically manipulated, or put in our crosshairs for sport or money."[10]

Regan opposes the utilitarian approach to thinking about what's right and wrong with respect to human treatment of animals "because it has no room for their equal inherent value or worth." Whereas Singer argues that sexual relations with an animal is morally permissible in certain circumstances, Regan counters that is categorically wrong. It is wrong because an animal is used merely as a means to human gratification, not as an end-in-itself fully entitled to the kind of treatment owed to a human being.

It is noted that Mary Ann Warren criticizes Regan for not providing an adequate explanation of why experiencing subjects of a life have inherent value. In her words: "Regan says that it is a postulate that subjects-of-a-life have inherent value, a postulate justified by the fact that it avoids certain absurdities which he thinks follow from a purely utilitarian theory. But why is it a postulate that subjects-of-a-life have inherent value?. ... If the reason is that subjects-of-a-life have an existence which can go better or worse for them, then why isn't the appropriate conclusion that all sentient beings have inherent value, since they would all seem to meet that condition?. ... If inherent value is based on some natural property, then why not try to identify that property and explain its moral significance."

There is merit in this criticism. Kant sought to provide a convincing rationale why humans have inherent worth—why rationality in humans is deserving of moral respect—but Regan provides no comparable explanation

10 Tom Regan, *Animal Rights, Human Wrongs*, Op. cit., pp. 65–66.

of what there is about nonhuman mammals that confers moral respect other than the broad generalization that they are experiencing subjects of a life.

Regan's views are also criticized for their uncompromising absolutism. As will be discussed in subsequent chapters, they are also criticized by biocentrists for failure to acknowledge what they (biocentrists) regard as the inherent value of all living beings, and by ecocentrists who contend that Regan's prescriptions are sharply at odds with management practices informed by conservation biology and ecology. Ecocentrists contend that vertebracentric thinking in general breaks down when questions such as the following are posed: If we have a duty to protect pets and domesticated animals from prey, do we also have a duty to protect animals in the wild from their prey? If overpopulation of a species is human-caused and very likely will result in the starvation of large numbers in the species population, is it morally wrong to selectively cull members to reduce overpopulation? More generally, does the good of the individual always take precedence over the collective whole of which the individual is a member?

ECOFEMINISTS AND OTHERS: CARING MATTERS MORALLY

David Hume and Adam Smith maintained that humans have a "natural sympathy implanted in our nature" (Hume's phrase). While these late-eighteenth-century thinkers believed this natural sympathy extended only to fellow humans, Arthur Schopenhauer contended that "the *animals* are also taken under its protection." Baird Callicott is a contemporary champion of the view that moral response issues from feelings of sympathy and compassion. Noting that Singer "heaped scorn on 'sentimental appeals to sympathy' ... and avowed that his animal welfare ethic was grounded exclusively in 'basic moral principles' ... in reason, not emotion," Callicott alludes to the alternative course taken by Hume, who "regarded altruistic feeling and sentiment as the edifice upon which ethics is erected," and by Darwin, "who explained how we came to have such feelings ... with reference to the evolutionary principle of natural selection."[11] Callicott discusses how this

11 Baird Callicott, "Environmental Liberation and Environmental Ethics: Back Together Again," in *Environmental Philosophy: From Animal Rights to Radical Ecology,* Michael Zimmerman et al., eds. (Upper Saddle River, NJ: Prentice Hall, 2005), p. 134.

sympathy has enlarged historically to encompass progressively larger communities, ranging from family to the global human community, and from domesticated animals to the all-encompassing biotic community.[12]

In addition to or in place of a justice approach, ecofeminists see emotional attachment and sympathetic caring as a more elemental basis for thinking about human responsibilities toward animals. In her book *Animals and Why they Matter*, Mary Midglely argues that social bondedness and the emotions connected with them are morally paramount, not reason alone.[13]

Writers in the feminist tradition view moral response as issuing in feelings of sympathetic care and sincere attempts to imagine the reality of another. "When my caring is directed to living things," Nel Hoddings writes, "I must consider their natures, ways of life, and needs. And although I can never accomplish it entirely, I try to apprehend the reality of the other."[14] Iris Murdock elaborates: "The direction of attention is outward, away from the self, which reduces all to a false unity, toward the great surprising variety of the world and the ability to so direct attention is love."[15]

Sympathetic caring is described in feminist literature as attentiveness to another's reality and as a sincere attempt to imagine how another feels. Moral imagination of this sort involves taking another's point of view and imagining not only how one would feel if one were in the other's place but how the other feels. Iris Murdoch uses the term "attentive love" to describe what is needed to sympathetically identify with another." Ecofeminists generally maintain that sympathetic care experientially and logically precedes justice—that caring dictates who are counted as morally relevant and to whom fair and humane treatment is owed. Caring that encompasses not only the pain and suffering of animals but also their yearnings, strivings, inclinations, and remarkable skills and accomplishments is all that is needed for recognition of

12 Callicott focuses on this enlarging sense of sympathetic identification as the key to bridging the chasm between animal rights and environmental ethics and these with anthropocentrism. He suggests that the "concentric rings" of what we care about spread outward, that each ring carries its own set of obligations, and that, generally, obligations associated with inner and more intimate rings outweigh obligations corresponding to the outer more global and inclusive rings.

13 Mary Midgley, *Animals and Why they Matter* (Athens, GA: University of Georgia Press, 1983). In contrast with Callicott's metaphorical comparison of caring to concentric circles with oneself in the middle, Midgley portrays them as overlapping concerns.

14 Nel Hoddings, *Caring: A Feminine Approach to Ethics and Moral Education* (Berkeley: University of California Press, 1984), p. 16.

15 Iris Murdoch, *The Sovereignty of Good* (1971), p. 66.

their worthiness, particularly when such caring is combined with the realization of the commonalities that humans share with animals and how we live in complex relations of interdependency. For the ecofeminist, care and the imputation of worthiness are mutually reinforcing and inseparable. We do not care about something seen to be worthless, but conversely, when we care, we see worthiness.

Many ecofeminists also emphasize that ethics exist in a political context. As Josephine Donovan put it, "People exercising attentive love see the tree, but they also see the logging industry. They see the downed cow in the slaughterhouse pen, but they also see the farming and dairy industry. They see the Silver Spring monkey, but they also see the drug corporations and university collaboration."[16] "Like many other feminists," Donovan says, "I contend that the dominant strain in contemporary ethics reflects a male bias toward rationality, defined as the construction of abstract universals that elide not just the personal, the contextual, and the emotional, but also the political components of an ethical issue. Like other feminists, particularly those in the 'caring' tradition, I believe that an alternative epistemology and ontology may be derived from women's historical, economic, and political practice."[17]

Brian Luke agrees that the justice arguments of Singer and Regan, while intellectually sound, are not enough. He argues that "direct sympathetic responsiveness to need is more central to animal liberation than concerns about consistency,"[18] and then goes on to discuss the myriad ways that industries in contemporary society undercut our gut-level capacity to respond to the suffering inflicted on animals in laboratories, factory farming, and sport hunting. He concludes with this summary: "The lesson I draw from this analysis is twofold, part heartening and part sobering. Heartening is the realization that the ethical basis of animal liberation is very simple and generally moving. A straight-forward presentation of what the animals are like and what is done to them ... can stir people, especially if the ideologies that block sympathy are simultaneously debunked. But sobering is a grasp of the nature of the social

16 Josephine Donovan, "Attention to Suffering: Sympathy as a Basis for Ethical Treatment of Animals," in *Animal Rights*, Josephine Donovan and Carol J. Adams, eds. (New York: Continuum, 1994), p. 150.

17 Ibid., p. 147.

18 Brian Luke, "Justice, Caring, and Animal Liberation," in *Beyond Animal Rights*, Josephine Donovan and Carol J. Adams, eds. (New York: Continuum, 1994), p. 99.

forces allied against a true perception of animals, against an understanding of what is done to them, against the possibility of acting from compassion. The substantial power of institutionalized animal exploitation sustains ignorance, promotes fear, rewards cruelty, and punishes kindness."[19]

Max Scheler, regarded as the founder of the phenomenological method in the social sciences, elevates sympathy to a form of understanding (*Verstehen*). Seeing the need for a counter-balance to objectification in science, he argued that investigators need to "redevelop a sympathetic capability to decode the symbolic language of nature." "We can," he says, "understand the experience of animals by attending to the behavioral and expressive signs [that reveal] an animal's emotional and psychological state." A dog's wagging its tail is a "grammar of expression" expressing joy. Similarly, other forms of life have a "grammar of expression" that humans can come to understand. To understand the language of other life, "We must," Scheler believes, "rid ourselves henceforward of our one-sided conception of Nature as a mere instrument of human domination. ... We must learn once more 'to look upon Nature as into the heart of a friend.'"[20] Indigenous people know this intuitively. Walking Buffalo, a member of the Stoney Tribe of Canada, reportedly said this about trees: "Did you know that trees talk? Well they do. They talk to each other, and they'll talk to you if you listen. Trouble is, white people don't listen. They never learned to listen to the Indians so I don't suppose they'll listen to the other voices in nature."[21]

THE ANIMAL ETHICS OF DAVID DEGRAZIA

In his book *Taking Animals Seriously*,[22] David DeGrazia develops a comprehensive and systematic animal ethic. In the introductory chapter of the book, DeGrazia states that his overarching goal is "to explore the mental life and moral status of animals in a philosophically penetrating, empirically

19 Ibid., p. 99–100.

20 For Rolston, an organism's inherent or intrinsic value is what makes it worthy of moral respect. For Taylor, respect emanates from an attitudinal orientation based on the beliefs that constitute the biocentric outlook.

21 Quoted in David Hughes, *American Indian Ecology* (El Paso: Texas Western Press, 1983), p. 49.

22 David DeGrazia, *Taking Animals Seriously: Mental Life and Moral Status*, (Cambridge, UK: Cambridge University Press, 2001). DeGrazia is also the author of *Animal Rights: A Very Short Introduction* (Oxford, UK: Oxford University Press, 2002).

well-informed way," and that, under this umbrella, the book has several more specific goals: "First, I want to transcend the utility-versus-rights debate and offer a well-developed methodology, a version of the coherence model, for fruitful pursuit of questions in animal ethics. ... Second, in exploring the mental life of animals, I strive for both philosophical rigor and empirical richness. ... Third, I endeavor to explore the basic features of animal well-being, especially in comparisons with humans."

The overall strategy involves examining basic ways that humans and other sentient animals can be harmed by humans and determining the respects in which each of these harms are comparably harmful to humans and nonhumans. Then, employing two basic moral principles—the principle of nonmaleficence and the principle of equal consideration—DeGrazia derives a practical set of guidelines or norms defining what constitutes morally right treatment of animals. DeGrazia's analysis of harm comparability is based on a detailed examination of what is known or can reasonably be inferred about the mental life of animals.

As a starting point, De Grazia characterizes what he calls the commonly held "phylogenetic scale" of moral status:

> Humans are placed at the very top of the [phylogenetic] scale, with Great Apes and dolphins a bit lower. Elephants, gibbons, and monkeys are somewhat lower, with canines and felines a bit lower, and rabbits and rodents lower still. ... Mammals are higher than birds, which are higher than reptiles and amphibians, which are higher than fish. ... As one moves down the phylogenetic scale, the amount of consideration one owes to beings at a particular level decreases. At some point one reaches beings that deserve just a little consideration. ... Beings below this line have no moral status.

Arguing that the "common view" is too crude but at the same time acknowledging that differences in moral significance is a sound intuition, part of DeGrazia's strategy is to clearly delineate the respects in which distinctions in moral significance are relevant and defensible.

The Mental Life of Humans and Other Animals

Based on what we know phenomenologically about our own experience of *pain* and the observed behavior of other animals, findings in neurophysiology and neuroanatomy, and functional-evolutionary hypotheses, DeGrazia concludes there are good and sufficient reasons for concluding that many animals feel pain.[23]

There can be little doubt also that many animals experience emotions as well, including *suffering* ("a highly unpleasant emotional state associated with more-than-minimal pain or distress"), *fear* (a type of distress defined as "a typically unpleasant emotional response to a perceived danger, usually in the immediate environment, a response that focuses attention to facilitate protective action"), and *anxiety* (another type of distress defined as "a typically unpleasant emotional response to a perceived threat to one's physical or psychological well-being, a response that generally inhibits action and involves heightened arousal and attention to the environment"). Other emotional states are common among animals as well, including *frustration* and *boredom.*

In response to the claim that language is necessary for rich emotional experience, DeGrazia points out that language is not required for *having* feelings, as opposed to verbally *expressing* those feelings. Human babies obviously do experience pain, pleasure, and fear even though they have not yet acquired a language enabling them to communicate their distress verbally. In examining the conditions of various commonly felt emotions and relationships among them, DeGrazia concludes that the capacity of animals to feel pain and other emotions strongly suggests that some animals have a sense of temporal self-awareness, of persisting as a self over time. "Fear would be impossible," DeGrazia says, "unless the subject has some awareness of

23 As noted by DeGrazia, Gary Varner identified six indicators that animals experience pain: (1) the presence of nociceptors (neural end organs that detect noxious stimuli); (2) possession of a central nervous system; (3) the connection of nociceptors to the central nervous system; (4) the presence of endogenous opioids (e.g., endorphins, enkephalins); (5) evidence that behavior in response to noxious stimuli is modified by analgesics; and (6) evidence that animal behavioral responses are analogous to human behavior responses. Varner concluded that mammals and birds satisfy all of these criteria, and that reptiles, amphibians, and fish satisfied most of them. Based on more recent data suggesting that reptiles and fish have nociceptors and are responsive to analgesics, DeGrazia contends it is safe to conclude that teleost fish and reptiles feel pain—and that amphibians do as well since they lie evolutionarily between fish and reptiles. The question whether cephalopods (squid and octopi) also feel pain remains an open question.

persisting into the future." Also, "there is considerable independent evidence that many vertebrate animals have memories and expectations for the future. Further, "the growing field of ethology—which examines the evolutionary origins animal behavior—tends to support the attribution of beliefs, desires, and intentional actions to many animals.

The central claim is that the *best explanation* of their behavior, given everything we know about them, requires these attributions. While DeGrazia believes the concept of personhood is too vague and contentious to be useful in animal ethics, he concludes that it is reasonable to attribute at least minimal degrees of intentional action, thinking, sociability, and self-awareness to mammals and certainly the higher primates ("even though such capacities may not meet the threshold for autonomy"), and that "these animals possess as much of these capacities as do many humans."

An important question for ethics is whether emotions experienced by humans are comparably experienced by other animals—whether suffering, fear, anxiety, loneliness, etc. are essentially the same for all animals or whether there are significant qualitative interspecies differences. Based on a review of the scientific and philosophical literature pertaining to this question, DeGrazia comes to this conclusion: "Since suffering is most fundamentally an experiential harm, it is reasonable to hold that if a human and an animal experience roughly equal amounts of suffering—however difficult, in particular cases, that may be to determine—they are comparably harmed."[24]

With regard to other negative emotions such as anxiety, fear, frustration, and loneliness—each is examined by DeGrazia in some detail—the conclusion is that there is not a close comparability in how they are experienced by humans and nonhumans. In regard to anxiety, for example, DeGrazia says this: "While the available evidence, taken together, supports the conclusion [that many animals experience anxiety], it does not imply that human anxiety and animal anxiety are qualitatively similar beyond a common unpleasantness and heightened arousal and attention. Undoubtedly, the language-laden complexity of human thought produces anxious experiences very different from those of animals. The present claim is that animals representing a wide range of species are capable of having anxious states, as captured in our definition of 'anxiety.'"[25]

24 DeGrazia, Ibid., p. 54.
25 DeGrazia, Ibid., pp. 47–48.

Harm and the Principles of Nonmaleficence and Equal Consideration

Two basic deontological principles are employed by DeGrazia in deriving a set of fifteen norms pertaining to the rightful treatment of animals: the principle of nonmaleficence and the principle of equal consideration. The former states that one ought not cause direct harm unnecessarily. The latter states that comparable harm ought to be given comparable consideration in moral deliberation. The principle of nonmaleficence is found in virtually all ethical systems and is, DeGrazua says, noncontroversial. He also thinks that the principle of equal consideration rests on sound intuition and that those who deny its validity have the burden of proving its inadequacy, which proof deniers have failed to provide.

Per the principle of nonmaleficence, it is fundamentally morally wrong to afflict avoidable harm on a sentient being, human or nonhuman. Based on the conclusion that humans and other sentient animals experience pain and the suffering that attends pain *comparably*, the principle of equal consideration requires that the suffering of humans and other animals be given equal consideration (regarded as equally moral significant) in moral deliberation.

Because the two moral principles employ the language of harm (rather than needs, preferences, or interests[26]), DeGrrazia develops a harm typology for use in the derivation of practical moral guidelines for animal treatment. In addition to inflicting pain and suffering, primary ways in which animals are harmed is killing, confining, and disabling them.

Because pain and the suffering it occasions are comparably felt, there is no basis for claiming that suffering should count any more or less morally than that of any other sentient being. In regard to *suffering*, It does not matter morally whether the harmed party is a human, a mammal, a bird, a fish, a reptile, an amphibian, or any other sentient animal. With regard to *death*, defined as foreclosure of opportunities for experiential well-being, DeGrazia sees death as harmful to all sentient beings, but not comparably harmful since human opportunities for experiential well-being are greater than other animals. In regard to *confinement,* DeGrazia's position is that such harm is comparable to the extent that it causes suffering, but not comparable with respect to widely varying species-typical functioning.

26 DeGrazia, as Singer, views an interest in not being harmed as the most basic interest one can have.

In summing up his analysis of how the several types of harm are, and are not, experienced comparably among sentient animals, DeGrazia said, "Our discussion has yielded several conclusions. First, a certain amount of suffering is a comparable harm, regardless of who the sufferer is. Second, when comparing humans with some animals—at least those 'below' mammals—death is not a comparable harm. In normal circumstances, death harms humans more. Third, confinement is a comparable harm across species in terms of causing a certain amount of suffering; it is not comparable, when comparing humans and at least some animals, in terms of interfering with valuable activities or possibilities for satisfaction."[27]

Rightful Animal Treatment: Fifteen Guiding Principles

Based on the two moral principles and his examination of the mental life of animals, DeGrazia then derives fifteen norms or principles (stated as injunctions, prohibitions, and related presumptions), which together provide practical guidelines for the rightful treatment of animals.

The first of the fifteen norms is a general statement of the principle of nonmaleficence: It is prima facie wrong to unnecessarily harm any being capable of being harmed. The second principle calls for reasonable efforts not to provide financial support for institutions that cause avoidable harm to sentient animals, including refusal to buy the products of factory farms. In regard to the harm of *suffering*, two obligations are derived: the injunction not to cause suffering solely for the sake of one's own or others' enjoyment, and the injunction to apply equally any standards allowing the cause of suffering. In regard to the harm of killing, the probation against killing humans, great apes, and dolphins is virtually absolute and of greater moral significance than the killing of fish, amphibians, and reptiles. In regard to the harm of *confining* (caging, chaining, isolating, fencing, and the like), two obligations pertain: the prohibition against unnecessary confinement and the injunction to ensure that any (justified) confinement is responsive to the animals' needs (providing ample space, light, food, exercise, companionship, and the like). In this connection, two additional presumptions are stated: the prohibition against confining innocent humans, great apes, and dolphins is virtually absolute and of greater moral significance than the confinement of other

27 DeGrazia, Ibid., p. 65.

non-dangerous sentient animals, including fish, birds, and herpetofauna. In regard to the harm of *disabling* a sentient animal (wing clipping, de-beaking, hobbling, and the like), he concludes that the prohibition against disabling humans, great apes, and dolphins is virtually absolute and that exceptions are justified only when disablement is necessary to avoid danger and a least-harm method of disablement is employed. In regard to the *keeping of pets*, two requirements are paramount: to provide for the pet's physical and psychological needs, and to enable a quality of life comparable to what the pet would have if it were not your pet. Finally, if there appears to be a conflict between benefiting an animal and respecting its autonomy, unless the former if very great and the latter is very slight, respect its autonomy.

DeGrazia contends that all but two of these normative guidelines are based on inferences drawn from the uncontroversial idea that it is wrong to inflict unnecessary harm (the principle of nonmaleficence), together with insights about the ways in which sentient animals can be harmed. In summarizing the overall derivation of the fifteen norms, DeGrazia states that "I have argued that equal consideration for animals is more reasonable than its denial, given the failure of opponents of equal consideration to meet their burden of proof. ... Anyhow, those who doubt that animals should be extended *equal* consideration should note that a commitment to giving animals *serious* consideration would be enough to support most of the conclusions."[28]

DeGrazia employs the fifteen moral norms to critique factory-farming practices and the capture and captivity of wild animals in zoos. In regard to factory farming, the general conclusion is that factory farms cause massive harm to millions of animals for trivial purposes and are ethically indefensible because they inflict extensive pain and suffering and otherwise harm these animals by unnecessarily confining, disabling, and killing them.

DeGrazia allows that a case can be made that the factory farming of chickens is less problematic morally than that of cattle and hogs, but he insists that "no plausible argument can be made that the fate of factory-farmed chickens is necessary, only that the way chickens are treated may be less bad.[29] In regard to the eating of fish, cephalopods, shrimp, crabs, and lobsters, DeGrazia comments that "Those who believe that these creatures are not harmed in death, and who doubt they suffer, might hold that the minor

28 Davis DeGrazia, *Taking Animals Seriously*, Op cit., p. 281.
29 DeGrazia, Ibid, p. 285.

harm of any short-term unpleasantness they may feel is compensated for by the gains for humans: health, convenience, pleasure, and enjoyment for fishermen." DeGrazia disagrees: "My thesis is that death harms the sentient, and my inclination to give fish and cephalopods the benefit of any doubt regarding suffering, lead me to a different view. ... But certainly, buying fish is much less morally problematic than buying factory farmed or even family-farmed meat".[30] Given the unsettled question whether shrimp, crabs, and lobsters are sentient, DeGrazia is personally disinclined to condemn the eating of these animals.

While DeGrazia contends that the ethical case against the eating of meat is based on requirements of the principles of maleficence and/or equal consideration, he cites other reasons why one ought not support factory farms financially or otherwise. His summary of these (teleological) reasons are: "First, factory farming has had a terrible impact on the environment in terms of pollution and the consumption and inefficient use of soil, energy, and water. Second, factory farming has a pernicious effect on the global distribution of food for humans. ... Third, factory farming in the United States has greatly harmed family farmers and rural communities by putting some three million farms out of business. ... Finally, meat eating, especially at average American rates, is associated with several health problems, [including] heart disease, stroke, diabetes, atherosclerosis, and certain cancers."[31]

THE ANIMAL WELFARE MOVEMENT

Laws to protect selected animals from abuse at the hands of humans gained momentum in England during the nineteenth century. Philosopher Jeremy Bentham contributed to the cause with this often-quoted statement that "the question is not, Can they *reason?* Nor, Can they *talk?* But, Can they *suffer?*" Prior to Bentham, few philosophers argued that the human treatment of animals is a matter of moral concern. A notable exception was Frenchman Jean-Jacques Rousseau who linked the treatment of animals to natural law. After Bentham and up to the mid-1900s, Arthur Schopenhauer was the only other major philosopher to address the question of the moral status of

30 DeGrazia, Ibid, p. 289.
31 DeGrazia, Ibid, p. 289.

animals. Noting that "Europeans are awakening more and more to a sense that beasts have rights," Shopenhauer chided Kantian ethics and Christian morality for "leaving animals out of account,"but applauded the English as "the first people who have, in downright earnest, extended the protecting arm of the law to animals."[32]

The first animal protection group in the US was the American Society for the Prevention of Cruelty to Animals (ASCPA), founded by Henry Bergh in 1866. Numerous national, regional, and local animal welfare organizations have since been established in both government and the nonprofit sector to promote kindness to animals, create public awareness, provide educational and information services, operate shelters and sanctuaries, provide first aid and other medical services for sick and injured animals, operate adoption services, and engage in political action. Some of the best-known organizations at the national level, in addition to ASCPA, include the Humane Society, People for the Ethical Treatment of Animals (PETA), Defenders of Animals, the Progressive Animal Welfare Society (PAWS), SPAY/USA, Voices for Pets, and the Animal Welfare Coalition.

Some animal welfare organizations are self-professed activists willing to engage in illegal activities to advance their causes. An example is the Animal Liberation Front, founded in 1976 by Ronnie Lee and Cliff Goodman. Operating as a grassroots guerilla organization "dedicated to the liberation of animals from all forms of cruelty and persecution at the hand of man," AFL activists see themselves as a modern Underground Railroad whose objective is to rescue animals from animal farms and laboratories and move them to sympathetic veterinarians and ultimately to sanctuaries. Threats, intimidation, and arson have eroded public support for the ALF in recent years, and the US Department of Homeland Security has put ALF on its list of domestic terrorists. Earth First! is another group using direct action and

32 Schopenhauer was no doubt alluding not only to Bentham in this regard, but also to protections afforded to animals by Parliament, which in 1882 adopted An Act to Prevent the Cruel and Improper Treatment of Cattle, with fines of up to five pounds or two months' imprisonment for persons who "beat, abuse, or ill-treat any horse, mare, gelding, mule, ass, ox, cow, heifer, steer, sheep or other cattle." The champion of this pioneering legislation in the House of Commons was Colonel Richard Martin, and it was Martin who brought the first prosecution. Martin was instrumental in establishing the Society for the Prevention of Cruelty to Animals in 1824 to monitor compliance with the 1822 act. In 1835, Parliament passed the Cruelty to Animals Act, legislation that amended the 1822 act and bought cock-fighting, baiting, and dog fighting under its jurisdiction.

sometimes illegal tactics. Formed in 1979, "not [as] an organization, but a movement," Earth First! believes in using all of the tools in the toolbox, from grassroots and legal organizing to civil disobedience and monkeywrenching." Still another example is the Sea Shepard Conservation Society, whose mission is "to end the destruction of habitat and slaughter of wildlife in the world's oceans."

Activist Henry Spira is a proponent of lawful nonviolent change, favoring what he calls "reintegrative shaming." His first campaign in 1976 successfully persuaded the American Museum of Natural History to abandon experimentation on cats. Perhaps his most notable achievement was "shaming" Revlon to stop using animals in cosmetics testing and to set up and contribute to the Center for Alternatives to Testing. Spira's approach has been widely adopted by other animal welfare groups, most notably the People for the Ethical Treatment of Animals. PETA members see themselves as "new welfarists" and "animal protectionists" rather than "animal rights" advocates.

Protection of animals in the legal system has reflected the growing chorus of voices calling for institutional and corporate change. The first legal code protecting animals in the US was The Body of Liberties adopted by the Massachusetts Bay Colony in 1641. One provision of this code states that "Noman shall exercise any Tirrany or Crueltie toward any bruite Creature which are usuallie kept for man's use." Many state and municipal legislatures have since enacted legislation protecting animals, and numerous regulatory bodies and national associations have adopted codes of conduct and certification requirements pertaining to the use of animals in research, factory farming, and the transportation of exotic and other animals in international and interstate commerce. Animal law is taught at law schools across the country, and the idea of establishing legal guardians to represent animals in courts of law is advocated by Christopher Stone and has the support of Alan Dershowitz and Laurence Tribe of the Harvard Law School.[33]

33 See Debbi Legge and Simon, *Law Relating to Animals* (Cavendish, 1997) for an overview of the development of legal protection for animals in the US and Canada.

7 BIOCENTRIC PERSPECTIVES

Vertebrate species comprise only about 4 four percent of all known species, and, if total population numbers within all species were tallied, the total number of living vertebrates would constitute a very small fraction of all living beings. Is it reasonable to assert that all non-vertebrates are morally considerable in their own right? What about invertebrate animals, or plants, or microbes? What about species or genomic kinds? Do any or all these biotic entities count morally? Are some or all of them the kinds of entities to which direct obligations could meaningfully be said to be owed?

This chapter explores how questions such as these have been addressed by biocentric ethicists. Beginning with a review of proposed biocentric criteria of moral standing, the discussion will revisit the question of whether analogues of human value and valuing are associated with the lives of nonhuman beings. Following this discussion, the question of the moral standing of species is addressed. Next, the systematic biocentric ethics of Paul Taylor is examined. The chapter then concludes with a brief review of legal protections extended to endangered species.

BIOCENTRIC CRITERIA OF MORAL STANDING

Albert Schweitzer wrote in the first half of the twentieth century that an ethics of reverence for life is rooted in the intuition that "I am life that wills to live in the midst of life that wills to live." For Schweitzer, to will something is to value it; willing to live is a valuing of life. A human intuitively knows that he or she wills to live, that one is *for* one's life, *for* one's existence and flourishing. This intuition is vivid when one's life is in peril: when one realizes that a misstep in a dangerous situation can lead to death, or that infirmity or dying means unfulfilled hopes and unfinished projects. One is also intuitively aware that other living beings will to live, that they, too, are *for* their lives. This awareness creates a sense of solidarity with other life; one sees the value of other lives in the same way as one sees the value of one's own life, as life willing to live, as life valuing life.[1] In compassionate persons, Schweitzer taught, this awareness blossoms into an attitude of reverence for life.

Kenneth Goodpaster wrote several decades later that "Nothing more than the condition of *being alive* seems to me to be a plausible and non-arbitrary criterion [of moral considerability."[2] Goodpaster sought to expose flaws in arguments limiting moral standing to humans, primarily by showing how the concept of moral considerability is confused with the concept of moral significance. He associated the condition of being alive with the capacity of living beings to perpetuate their lives through organization informed by homeostatic feedback. "The core of moral concern," he wrote, "lies in respect for self-sustaining organization and integration in the face of pressures toward high entropy."[3] Lawrence Johnson similarly characterized "being alive" as "a persistent state of low entropy sustained by metabolic processes for accumulating energy [wherein] organic unity and self-identity are maintained in equilibrium by homeostatic feedback processes."[4]

Other writers have used the term "self-actualization" to describe what it is about living organisms that makes them morally considerable in their own right, whereby the phenomenon of self-actualization is variously characterized as the capacity of organisms to actualize biological potential; to progress developmentally toward end-state or goods of their kind; to adapt in ways that enable them to thrive in oftentimes perilous environments; or, as

---

1 The interpretation of what Schweitzer means by "will to live" is based on discussions the author had with Dr. Schweitzer in Lambarene, Gabon, in 1964. The quotation is from Schweitzer's *Teaching of Reverence for Life* (Holt, Rinehart, & Winston, 1965), p. 26.

2 Kenneth Goodpaster, "On Being Morally Considerable," in *Journal of Philosophy*, 75. Goodpaster's article is included in *Environmental Ethics: Readings in Theory and Practice*, Louis P. Pojman, ed. (Wadsworth, 2001).

3 Kenneth Goodpaster, "Moral Considerability," in *The Journal of Philosophy*, 75, p. 323.

4 Lawrence Johnson, "Toward the Moral Considerability of Species and Ecosystems," in *Environmental Ethics* 149, 1992. See also Johnson's *A Morally Deep World* (New York: Cambridge University Press, 1991).

Holmes Rolston puts it, to be "value-able" in these ways. Self-actualization is seen as a good that is intrinsic to (inherent in) the nature of a living being, and as a phenomenon that constitutes the good of a living being. Realization or actualization of potentiality is recognized as a basic value in human life, and there is no good reason, biocentric thinkers contend, to deny that self-actualization in other life has comparable value.

Many believe that the granting of moral status to invertebrate animals, plants, and microbes flies in the face of common sense. On a planet with millions of species and countless trillions of living organisms, critics say it is confusing, if not maddening, to be told that all of them are morally considerable in their own right. When organisms feed on other organisms throughout nature and when numerous organisms pose threats to human health and safety, it is absurd, they maintain, to suggest that all of these trillions upon trillions of organisms deserve moral consideration. Why worry about a virus that makes us sick? Why fuss over snail darters when people need electricity? Questions such as these typically focus on pests, pathogens, or microbes with huge populations as part of a *reduction ad absurdum* strategy to shut down further discussion. Goodpaster was among the first biocentric thinkers to point out that this strategy conflates the concepts of moral standing and moral significance. The recognition of an entity's moral standing is a recognition that harm to the entity is a morally relevant question; it is not a claim that it is wrong to do what is necessary to protect human health and safety. If a population of microbes is judged to be a serious threat to human health and not an essential component of well-being in some greater scheme of life, the obviously morally justifiable course of action would be to eliminate the threat.

Some writers have claimed that Schweitzer's use of the term "will to live" or Paul Taylor's characterization of an organism as a "teleological center of life" are anthropomorphic and impute human characteristics such as "striving," "desiring," "aiming," and "goal-seeking" to life that lacks the mental or cognitive abilities that these terms connote. But a close reading of these biocentric theorists suggests that this line of criticism is not plausible. In explaining what self-actualization in an organism involves, Goodpaster, Johnson, Taylor, Rolston, and others provide naturalistic explanations that are not anthropomorphic. Also, when a chaos-complexity theorist posits that "attractors" (values) function in complex systems as trajectories toward which the system as a whole tends and without which the system tends toward disintegration and collapse, the explanation is not framed in anthropomorphic terms but rather makes reference to a hypothesized fundamental feature of all self-organizing biotic entities in nature.

Seeing all life in human-like terms needs to be guarded against, and important differences between humans and nonhumans need to be recognized. But humans share important commonalities with many life forms past and present, the most basic of which is the capacity to self-organize. Although only humans ask value questions and make value judgments, humans are not the only entities capable of tracking and realizing valued end-states. And while the cognitive abilities of humans far surpass those of other animals and certainly plants and microbes, nonhuman life is not utterly devoid of intelligence. The capacity of even the simplest organisms to develop and reproduce is rooted in an information-processing intelligence encoded in their DNA.

HOLMES ROLSTON: REVELATIONS OF NATURAL VALUE

Holmes Rolston developed a naturalistic theory of value that synthesized insights from numerous disciplines within the Darwinian paradigm. Central to his thinking is the thesis that value permeates nature. It is displayed in organisms, species, and ecosystems, and is "on show," he says, for humans to contemplate and understand. In expounding upon this thesis, several ideas recur in his writings: that the evolution of life on planet Earth reveals adaptive intelligence, not only natural selection through random variation; that evolutionary development has produced a vast profusion of life-forms in ever-increasing levels of diversity, complexity, organization, and intelligence, including the human brain and its creation, culture; that, while many players have participated in life's immense journey, only humans are capable of telling the story of their participation; that those who contemplate life's three-billion-year-old history and themselves participate in the telling of its story cannot but give up the idea that value and moral worthiness are confined to human culture; and that many stories, consistent with the Darwinian story line, are needed to account for all the creativity, innovation, and valuational thrust.

In his article "Value in Nature and the Nature of Value,"[5] Rolston describes the value that is "on display" in wild animals, in genomic expression, in ecosystems, and on a planet where life has evolved over several billions of years. At the conclusion of this article, Rolston revisits the claim that there can be no value without a valuer. "The problem with [this] axiom" he says, is "that it is too individualistic; it looks for some center of value located in a subjective self." "But," he continues, "that is not the whole account of value in a more holistic, systemic, ecological, global account." While perceptions require perceivers, religious beliefs require believers, and science requires scientists, there can be law without a lawgiver, history without an historian, and creation without a creator.

It takes a human to tell life's story and its many substories: how, for example, value is reflected in the caring of wild animals for their lives, how information-processing intelligence is involved in the genetically informed growth of a plant, or how evolution tends toward increasing levels of intelligence, organization, and awareness of self and environment. And, while it takes a human to set up the axiological scales with which value in nature can be revealed and assessed, the scales themselves do not create what is being evaluated, no more so than do quantitative measurement scales create what can thereby be measured.

Rolston acknowledges, indeed emphasizes, the differences between human and nonhuman life. "Humans superimpose cultures on wild nature out of which they emerged, with radical innovations," he says. "Information in wild nature travels intergenerationally on genes; information in culture travels neutrally as people are educated into transmissible cultures."[6]

But Rolston refuses to say that caring, awareness, and intelligence are unique to human life and culture. These capacities and their manifestations are disclosed throughout nature, present and past—but, again, with notable differences. Awareness in a single-cell organism is simple chemical awareness; in humans, it is self-conscious awareness associated with a highly complex system of electrochemical neural networks regulated in a brain

5 Holmes Rolston, "Value in Nature and the Nature of Value," in *Philosophy and the Natural Environment*, Robin Attfield and Andrew Belsey, eds. (Cambridge: Cambridge University Press, 1994), pp. 13–30. The article was presented at the Royal Institute of Philosophy's annual conference in 1993 and also published in *Institute of Philosophy Supplement: 36*.

6 Holmes Rolston, *Conserving Natural Value* (New York: Columbia University Press, 1994), p. 2.

composed of trillions of cells. Intelligence in a single-cell organism is associated with information-processing capability encoded in the DNA of its genome; in humans, intelligence is associated not only with the processing of information encoded in more complex sequences of DNA but with a brain specialized and complex enough to permit the kind of information processing involved in reasoning and symbolic communication and, with these, the creation and transmission of culture. Caring in its most elemental form is found in enzymatic self-reproduction and survival instinct. In its human and most universal form, caring transcends self-interest and takes into consideration the well-being of all life on the planet.

Rolston believes that it is morally naïve, if not morally perverse, to claim that only one among ten or more million species counts morally. In his words:

> Seven billion years of creative toil, several million species of teeming life have been handed over to the care of this late-coming species in which mind has flowered and morals have emerged. Ought not those of this sole moral species to do something less self-interested than count all the produce of an evolutionary ecosystem as rivets in their spaceship, resources in their larder, laboratory materials, recreation for their ride? Such an attitude hardly seems biologically informed, much less ethically adequate. Its logic is too provincial for moral humanity. Or, in a biologist's term, it is ridiculously territorial.[7]

THE MORAL STANDING OF SPECIES

Are species nothing more than labels used in biological classification, in which case any debate concerning their moral status would be pointless? Or are they discrete biological entities in the natural world, as real as the individuals that instantiate them, and thus the kinds of entities to which moral standing might reasonably be ascribed?

7 Holmes Rolston, "Duties to Endangered Species," in *Earth Ethics: Introductory Readings in Animal Rights and Environmental Ethics*, James Sterba, ed. (Prentice Hall, 2000, p. 328).

Species as Discrete Biological Entities

For biocentrist Paul Taylor, species are nothing more than labels used in biological classification. When we say that species change—that new species are created, or that existing species split into other species, or go extinct—what this means, Taylor maintains, is that changes occur in the populations of the named groups, changes that sometimes result in new names and new biological classifications. Taylor's view of species was held by Charles Darwin himself: "I look at the term species, as one arbitrarily given for the sake of convenience to a set of individuals closely resembling each other." Biologist A. B. Shaw agrees: "The species concept is entirely subjective."[8] For these writers, a species is no more real than a family, a genus, a phylum, a kingdom, or any other higher-order biological classification.

Other biologists, however, see species as having objective referents. G. G. Simpson states that species "are real biological entities in the natural world."[9] Viewing species as "interbreeding natural populations that are reproductively isolated from other such groups," E. Mayr states that "species are real units of evolution."[10] Viewing species as "entities that exhibit a pattern of phylogenetic ancestry and descent among units of like kind," Elderedge and Cracraft characterize species as "discrete entities in time as well as space."[11] For Stephen Gould, "Species are nature's objective packages."[12] In his view, evolution is like a bush, not a ladder, and species are like new branches on the bush that emerge quickly in comparison with the millions of years that many species populations have lived with only minimal change. In Gould's words: "The principle of bushes and the speed of branching" reveal that "continuous evolution can and does yield a world in which the vast majority of species are separate from all others and clearly definable at any moment in time."[13]

---

8 A. B. Shaw, "Adam and Eve, Paleontology, and the Non-Objective Arts," in *Journal of Paleontology* 43, pp. 1085–1098.

9 G. G. Simpson, *Principles of Animal Taxonomy* (New York: Columbia University Press, 1961), p. 153.

10 E. Mayr, "The Biological Meaning of Species," in the *Biological Journal of the Linnaeus Society* 1, pp. 311–320.

11 N. Eldridge and J. Cracraft, *Phylogenetic Patterns and the Evolutionary Process* (New York: Columbia University Press, 1980), p. 92.

12 Stephen Jay Gould, "What is a Species?" in *The Environmental Ethics & Policy Book*, Donald VanDeVeer and Christine Pierce, eds. (Thomson Wadsworth, 2003), p. 467.

13 Ibid., p. 467.

Holmes Rolston sits squarely in the "realist" camp. He describes a species as "a living historical form, propagated in individual organisms, that flows dynamically over generations" and as "a coherent, ongoing form of life expressed in organisms, encoded in gene flow, and shaped by the environment."[14] Higher order biological classifications do not have objective referents in the way that species do; unlike species, they do not refer to concrete biological entities that maintain an identity over time. For Rolston, "The assertion that there are specific forms of life historically maintained in their environments over time seems as certain as anything else we believe about the empirical world."[15]

Genomic Kinds and Moral Standing

Assuming that species are discrete biological entities in the natural world, no less real than the individuals that comprise and exemplify them, are they also the kind of biotic entities to which moral respect is due? An affirmative answer is supported by viewing species as genomic kinds.

The genomes of organisms having eukaryotic cell organization ranges from two to six hundred gene-carrying chromosomes, each in a combination of DNA, RNA, and protein within a membrane-bound nucleus. There is a basic division of labor in each cell: genes give instructions for the performance of cellular tasks, and proteins carry out these instructions. Each type of protein is a specialist that performs a single task; if a cell needs to do something faster or slower, it makes more or less of the protein; or, if it needs to perform a new task, it makes a new protein. Each protein is composed of a chain of amino acid molecules (DNA), and each DNA molecule is made up of four types of nucleotides (A, T, G, C). The base pairings and sequencing of nucleotides along the amino acid chain determine what the protein will do. Genetic instruction (expression) occurs when a gene is "read" by a cell, a process that involves transcribing a gene's DNA sequence into a very similar molecule called RNA, each exhibiting the same sequence of paired nucleotides, and then joining the two amino acid chains together to make a complete "folded up" protein chain. A similar process occurs each time a cell divides. Because

14 Holmes Rolston, "Duties to Endangered Species," Op. cit., p. 321.

15 Holmes Rolston, "Environmental Ethics: Values in and Duties to the Natural World," Op. cit., p. 74.

DNA is made of two strands with paired-up nucleotides holding the two strands together, the task of enzymes is to "unzip" the old strands, pair up new nucleotide units, and "zip" the strands back together again. The new unit contains one strand from the old DNA and one newly made strand. The process sometimes produces mistakes in the paring of nucleotides (e.g., G is paired with A), causing a mutation in gene sequence. Along with natural selection, mutations play an important role in evolutionary development.[16]

What is said to make a genomic kind morally considerable in its own right is its capacity to control growth and development over the lifetimes of individuals in which it is imprinted and to enable the transmission of this genetic blueprint intergenerationally. Genomic expression makes self-actualization in an organism possible, just as its intergenerational transmission through reproduction makes species perpetuation possible. Mutations play a role in this process, but as complexity theorist Stuart Kauffman hypothesizes, so does an underlying process of self-organization. Genomic expression is a valuing process involving information-processing intelligence. Surely the self-organizing capacity of a genome to regulate cellular activity and to enable its transmission to progeny is something worthy of moral respect.

Extinction is an ultimate harm. In Holmes Rolston's view, "Every extinction is a kind of superkilling. It kills forms, beyond individuals. It kills 'essences' beyond 'existences.'... It kills collectively, not just distributively. ... A shutdown of the life stream is the most destructive event possible." He continues: "If, in this world of uncertain moral convictions, it makes any sense to claim that one ought not to kill individuals, without justification, it makes more sense to claim that one ought not to superkill the species, without justification."[17] If killing a human being matters morally, then doing what could lead to the extinction of *Homo sapiens* certainly matters morally. The same logic applies to nonhuman life: if killing an animal matters morally,

16 In each of the approximately one hundred trillion cells in the human body, more than fifty thousand genes are carried on twenty-three chromosome pairs, with the entire human genome consisting of about three billion nucleotides. It is thought that only about 2 percent of this complex (its "exome") is protein coding; the other 98 percent of nucleotides are "switches" regulating when and where genes are expressed. According to the National Institutes of Health, about 85 percent of disease-causing mutations occur in the exome.

17 Holmes Rolston, "Challenges in Environmental Ethics," in *Environmental Philosophy: From Animal Rights to Radical Ecology*, Michael E. Zimmerman, J. Baird Callicott, George Sessions, Karen J. Warren, and John Clark, eds. (Prentice Hall, 2005), p. 136.

killing the last remaining members of its kind matters morally. A shutdown of a life stream without justification is a heinous act.

THE BIOCENTRIC ETHICS OF PAUL TAYLOR

Paul Taylor developed a systematic biocentric ethics in his book, *Respect for Nature*.[18] In this work, he defines what he calls the "biocentric outlook," which, he explains, engenders an attitude of respect for nature. He goes on to say that people who give their assent to the beliefs defining the biocentric will readily acknowledge the inherent worth of all living beings and will, moreover, grant the moral legitimacy of four basic prima facie duties that humans owe to other life. After identifying and explaining these prima facie duties, Taylor articulates five procedural principles for resolving the many conflicts that arise between human and nonhuman goods.

The Biocentric Outlook

For Taylor, an organism's worthiness of moral respect is a judgment made by a person who takes on what he (Taylor) calls the "biocentric outlook": an outlook defined by and grounded in four basic beliefs: that humans are latecomers in life's history on the planet but must be viewed as an integral part of an evolutionary process that began several billion years ago; that all living beings exist in systems interdependence in which the survival and well-being of each is determined not only by physical environmental conditions but also its relationship to other living things; that all organisms are teleological centers of activity in the sense that each is a unique individual exemplifying the functions and activities of its species and pursuing its own good in its own way; and that human beings are not inherently superior to other living beings.

Taylor explains what he means by calling organisms "teleological centers" in this way: "To say that an organism is a teleological center of life is to say that its internal functioning as well as its external activities … have the constant tendency to maintain the organism's existence through time and to enable it successfully to perform those biological operations whereby it reproduces its kind and continually adapts to changing environmental

18 Paul Taylor, *Respect for Nature* (Princeton, NJ: Princeton University Press, 1986).

events and conditions. It is the coherence and unity of these functions of an organism, all directed toward the realization of its good, that makes it one teleological center of activity."[19]

Taylor defends the belief that humans are not inherently superior to other living beings with two lines of argument. First, he argues that the first three stated beliefs are widely accepted scientific facts that discredit the claim of human superiority. Second, the claim of human superiority presupposes what is being claimed. In Taylor's words: "To use standards based on human value is already to commit oneself to holding that humans are superior to non-humans."

Persons who adopt or give their assent to these beliefs will readily attribute inherent worth to all entities having a good of their own and will, moreover, readily affirm the principle of the equal inherent worth of all living beings.[20]

Four Prima Facie Moral Duties

Taylor argues that a person who takes on the biocentric outlook will readily agree that humans owe four general duties to nonhuman life. While obviously not exceptionless, duties owed to nonhumans are nevertheless recognized to be binding by persons who share a biocentric outlook on life. In brief, the four *prima facie* duties are:

- **The duty of non-maleficence:** the duty not to unnecessarily harm a being that has a good of its own.
- **The duty of noninterference:** the duty not to interfere with processes essential to the good (teleological development) of a living individual.
- **The duty of fidelity:** the duty not to deceive or break the human-created trust of an animal, particularly when the trust is deliberatively created with the intent to kill or capture.[21]
- **The duty of restitutive justice:** the duty to make appropriate restitution for violations of the other three general duties.

---

19 Taylor, Ibid., pp 80–81.

20 For Rolston, an organism's inherent or intrinsic value is what makes it worthy of moral respect. For Taylor, respect emanates from an attitudinal orientation based on beliefs that constitute the biocentric outlook.

21 Examples include leaving food to subsequently lure an animal to a bated trap, or use of food, calls, or decoys to lure an animal within shooting range.

Although Taylor does not assign relative weights to these *prima facie* duties, he offers two general rules of thumb to guide their application: first, fidelity and restitutive justice generally override noninterference, particularly in cases in which interference does not cause permanent harm; and, second, restitutive justice generally overrides fidelity, particularly when no serious harm is done to a creature whose trust is betrayed.

Five Procedural Principles

When unavoidable conflicts between human and nonhuman interests arise (e.g., clear-cutting a woodland to build a medical center, destroying a freshwater ecosystem to build a lakeside resort, replacing a cactus desert with a housing development), Taylor offers five procedural principles for resolving such conflicts. Briefly stated, the five principles are:[22]

- **The principle of self-defense:** If necessary to protect human health and safety and the threat of harm is imminent and clear, it is permissible to kill, confine, or disable a threatening animal, plant, or microbe.
- **The principle of proportionality:** It is not morally permissible to sacrifice *basic* nonhuman needs (survival needs) for the sake of *nonbasic* human needs.
- **The principle of minimum wrong:** Human projects deemed necessary to achieve highly valued social goods are morally permissible only if no alternative means for achieving those goods involve fewer violations of prima facie duty.
- **The principle of distributive justice:** When a scarce resource is critically needed by humans and nonhumans alike, all needing parties should be allocated shares sufficient for their survival and basic needs fulfillment.
- **The principle of restitutive justice:** Restitution should be made for violations of prima facie duty (impermissible harm) to nonhuman life, provided that the good of restitution is greater than the harm compensated for. If restitution cannot be made to the nonhuman individuals

22 The statement of these principles are condensed paraphrases of Taylor's longer exposition.

harmed, restitution ought to be made to other members of their species in the form of habitat or other forms of protection.[23]

Criticisms of Taylor

Various criticisms have been leveled at Taylor's biocentric ethics. One fundamental criticism is that the principle of minimum wrong (allowing that a highly valued social good justifies violation of prima facie duty to nonhumans) contradicts Taylor's claim that all living beings have equal intrinsic worth, the cornerstone of his biocentric egalitarianism.

Some critics suggest that this inconsistency arises from Taylor not wanting his system of ethics to require so much of humans that it would be summarily rejected as too extreme and impractical. Whether or not this was Taylor's motivation, critics contend that Taylor's biocentric principles are, indeed, to extreme and impractical. The principle of distributive justice—which requires that human and nonhuman parties be given equal shares of a comparably needed resource—is singled out in this regard. Application of this principle would, it is argued, radically hamstring activities essential to human well-being, and that an ethics with this consequence cannot be taken seriously. Another concern related to the "impracticality" of Taylor's ethics is that the delineation of basic and nonbasic needs—required by the principle of proportionality—is difficult to make in practice.

Another criticism is that Taylor's biocentric *individualism* fails to recognize the central role and moral significance of collective wholes in nature: species and ecosystems. This failure leads to an inability, the criticism goes, to make morally significant distinctions between domesticated and wild animals, between abundant and endangered species, and between invasive and endemic members of ecosystems. These are questions of moral significance that expose problems for an ethical perspective focused exclusively on the well-being of individual organisms. While Taylor's ethic condemns practices

---

23 When harms are human-caused, restitution might involve one or more of the following approaches: *restoring* land to conditions similar to those that existed before they were modified (e.g., removing a dam); *repairing, replacing, or providing functional substitutes for* what has been damaged; providing timely *emergency aid* to head off imminent death, extinction, or destruction, including the administration of medical aid for life-threatening injuries; and *in-kind compensation*: preventing injury or death to other members of the species, or restoring a comparably damaged ecosystem.

that destroy habitats and indiscriminately interfere with natural processes, such obligations are derived from duties owed to individuals within ecosystems. The ecocentric rejoinder is that what benefits individuals (parts) may not be good for the ecosystem (the whole), and that, from a resource management perspective, the good of wholes may on occasion take precedence over the good of constituent individual members. In regard to species, Taylor denies their ontological reality, contending they are simply labels used to group populations sharing certain defined characteristics. As noted previously, Holmes Rolston and others argue that species are nature's evolved "objective packages" and are no less real than the individuals that instantiate them. For these writers, an ethics that fails to recognize the independent moral standing of species is seriously flawed.

PROTECTION OF ENDANGERED SPECIES

The alarm expressed by scientists, ethicists, and concerned citizens over accelerating extinction was discussed in Chapter 2. As noted there, it is estimated by credible sources that half of all species existing at the turn of the last century will have become extinct by the middle of this century. Wilcove and Master estimate that the "known species threatened with extinction is ten times higher than the number protected under the Endangered Species."[24] As discussed previously, various factors contribute to species extinctions, including habitat destruction pollution, disease, resource overexploitation, and climate change. But whatever weights might be assigned to causes in particular cases, what distinguishes the current wave of extinctions from those of the geological past is that the Holocene involves many plant and microbial species, not animal species primarily.

The stated purpose of the Endangered Species Act (ESA) is "to protect critically imperiled species from extinction as a consequence of economic growth and development untempered by adequate concern and conservation." ESA is administered by two federal agencies: the US Fish and Wildlife Service (FWS) in the Department of the Interior and the National Marine Fisheries Service (NMFS) in the National Oceanic and Atmospheric Administration.

24 D. S. Wilcove and L .L. Master, "How Many Endangered Species are there in the United States," in *Frontiers in Ecology and the Environment*, 3(8), pp. 414–420.

FWS is responsible for protecting listed terrestrial animals and freshwater fish; NMFS is responsible for protecting listed marine fish and wildlife.

A species can be "listed" as endangered or threatened in two ways: through the Candidate Assessment Program established by each of the two administrative agencies (FWS or NMFS) or by petition by an outside individual or organization. A "candidate" can be a taxonomic species, a subspecies, or, in the case of vertebrates, a "distinct population segment." Once a petition for listing is received, a review is conducted by the responsible agency, which then issues a judgment as to whether the petition is "not warranted," "warranted," or "warranted but precluded."

In regard to the listing process, Section 4 of ESA requires FWS and NMFS to designate specific areas as "critical habitat zones" on both public and private lands. An amendment to ESA in 1978 requires the two administrative agencies to list species "on the basis of the best scientific data available after taking into consideration the economic impact." When it became evident that the economic assessment provision had virtually stopped new listings and resulted in the withdrawal of nearly two thousand species for which petitions had been made, another amendment to ESA in 1982 eliminated the economic assessment requirement.

Section 7(a)(2) of ESA prohibits federal agencies from authorizing and funding projects that destroy or adversely modify critical habitats. While this provision applies to federal projects, large-scale private projects are brought under the umbrella of Section 7(a)(2) since they generally require federal permits. More than half of listed critical habitats to date are on non-federal property.

Both FWS and NMFS are required to formulate an Endangered Species Recovery Plan outlining the goals, tasks, costs, and estimate of the time required for saving a listed species. While no time limit was imposed for the completion of a recovery plan, actual completion times have taken considerably longer that the three-year policy target established by FWS. Recovery plans are themselves ranked, with the highest priority given to species most likely to be saved, particularly when the threat of extinction is attributable to construction or other development. An amendment to ESA in 1988 defined more precisely the information requirements of a recovery plan and added a public participation component.

A 1982 amendment requires that permits for "incidental takings" of an endangered species can be granted only if those "takings" are part of an approved Habitat Conservation Plan. A permit is given for a specified period of time and may be revoked at any time. Robert Thornton characterized habitat conservation plans as "a major force for wildlife conservation and a major headache to the development community."[25]

Still more amendments to ESA were made during the latter half of the 1990s. One of these amendments instituted the "no surprises rule" to protect private landowners if unforeseen circumstances make it impossible to achieve the levels of protection or mitigation specified in an approved habitat conservation plan. This rule absolves private owners of liability for setting aside additional land as may subsequently be required. Other amendments authorized the "safe harbor agreement" and the "candidate conservation agreement" programs. Under the former, a private party may agree to alter owned land in ways that enhance the survival of a listed species in exchange for assurances that FWS will permit future "takes" above a level specified in the "safe harbor" agreement. Under a "candidate conservation agreement," a private owner who agrees to restore, enhance, or maintain a critical habitat known to be in decline will not be required to undertake additional protection measures in the event that other species associated in the identified critical habitat are subsequently listed as threatened or endangered. While free market economic theorists generally laud these incentives to private land owners and believe additional incentives are needed to improve ESA's effectiveness, environmentalists see them as diluting the intent of the act and as unwarranted concessions to corporate interests.

Penalties for violations of provisions of the ESA can be fines as high as $50,000 and/or one-year imprisonment and civil penalties up to $25,000 per violation. The most punishable offenses are trafficking of endangered species and unauthorized "takings." Rewards can be paid for information leading to a conviction or revocation of a license or permit.

As of February 2012, 593 animals and 794 plants were listed as endangered or threatened. Recovery Plans were in place for nearly 80 percent of listed endangered species and about this same percentage of threatened species. From 1967 to 2012, fifty-one species were delisted, of which twenty-three had recovered, ten had gone extinct, ten had been taxonomically reclassified, and six had been discovered to have larger populations than

25 Robert D. Thornton, "Searching for Consensus and Predictability: Habitat Conservation Planning under the Endangered Species Act of 1973" in *Environmental Law*, 1991.

previously estimated. Notable examples of delisted species are the bald eagle, whooping crane, peregrine falcon, gray wolf, gray whale, grizzly bear, and the San Clemente Indian paintbrush. Twenty-five species had been down-listed from endangered to threatened.[26]

On the international front, the International Union for the Conservation of Nature (IUCN) has established a network of conservation organizations, scientists, and research stations around the world for the purpose of identifying and categorizing the conservation status of animals, plants, and other organisms. Nearly two hundred countries have signed an accord under the IUCN umbrella to create biodiversity action plans to protect imperiled species using the following classifications: Extinct, Extinct in the Wild, Critically Endangered, Endangered, Vulnerable, and Least Concern. IUCN periodically releases a "Red List" of endangered species.

NatureServe is an IUCN Red List partner that focuses its efforts in the US, Canada, Latin America, and the Caribbean. It uses a 1–5 ranking scale (1 = Critically Imperiled, 2 = Imperiled, 3 = Vulnerable, 4 = Apparently Secure, and 5 = Secure) to assess the conservation of species on three geographic scales: global (G), national (N), and state/province (S). The letter designations X and H are also used to classify species that are presumed to be extinct (X) and possibly extinct (H). NatureServe's database and classification system are widely used by government agencies, nongovernmental organizations, and researchers to monitor conservation status of known species in the region.

26 Current statistics are available online at http://ecos.fws.gov/tess/pub/Boxscore.do.

8 ECOCENTRIC PERSPECTIVES

Beginning with a brief review of the emerging science of ecology during the first half of the twentieth century, this chapter examines the ecosystem-oriented ethic of Aldo Leopold.[1] Differing interpretations of Leopold's writings are considered, along with various criticisms that have been leveled at Leopold's ecological approach to ethics. The question of the moral standing of ecosystems is then taken up. The last section of the chapter reviews the divergent views that have developed in conservation biology and among the many groups working in the field to conserve resources and preserve the richness of life that ecosystems harbor and support.

PIONEERS IN THE DEVELOPMENT OF ECOLOGY

Before publication of Darwin's *Origin of Species*, the number of species and their populations were generally assumed to be more or less fixed. It was recognized that changes occur in the natural world, of course, but the fossil record provided scant evidence before Darwin that life has been evolving for millions of years and that, during this immense expanse of time, the vast preponderance of species that have come into existence have since disappeared from the scene. In the wake of Darwin's observations of the great diversity of plants and animals around the world and questions he posed about their adaptations and ancestry—observations and questions that led to his evolutionary theory of natural selection—promoted

1 The terms "ecosystem-oriented," "ecology-oriented," and "ecocentric" are used interchangeably.

widespread interest in developing a fuller understanding of how diverse organisms in diverse natural environments relate to one another and to their natural environments. The German biologist Ernest Haekel coined the term "ecology"[2] to describe the emerging new science, with its emphasis on relationships, change, and the need for system-oriented and multidisciplinary thinking.

Frederick Clements, a pioneering ecologist in America during the early decades of the twentieth century, introduced the concept of succession to explain orderly biological changes in midwestern prairies and grasslands. On his account, succession begins with the development of a barren site and progresses through several stages: arrival of colonizing plant species, establishment of initial vegetation, competition of established species for space, light and nutrients, replacement of one plant community by another, and, finally, development of a climax community in which equilibrium or a steady state is reached. In the climax stage of development, the composition of species remains relatively stable and species populations are well adapted to climate and other environmental conditions. Clements saw an ecosystem as a kind of superorganism,[3] and, accordingly, saw the ecologist as akin to a physician: as the doctor relies upon anatomy and physiology to diagnose maladies and restore health to the human body, so the ecologist consults and integrates multiple scientific disciplines to diagnose and prescribe treatments to ensure healthy and balanced biotic communities.

In the 1920s, Charles Elton conducted studies into "nature's economy" or "nature's household" by analyzing the function that various organisms play in the food chain of a particular ecological community. The commodity that is exchanged in this economy is food, and its participants ("niche players") are food producers, food consumers, and food decomposers. Photosynthesizing plants and cyanobacteria are the producers (autotrophs); herbivores, carnivores, and omnivores are primary, secondary, and tertiary consumers (heterotrophs); and bacteria, fungi, and insects are the decomposers (detrivores) that restore nutrients to the environment and begin the trophic cycle anew.

2 The German word for "ecology" is "oekologie." The term "ecology" combines two Greek words: *oikos* ("home" or "household") and *logos* ("study of").

3 In Clement's words, "As an organism, the formation arises, grows, matures, and dies. … The climax formation is an adult organism, the fully developed community, of which all initial and medial stages are but stages of development. Succession is the process of the reproduction of a formation."

Clements observed that loss of a species can result in substantially altered trophic relationships throughout a food chain, sometimes with massive die-offs.

Englishmen Arthur Tansley is credited for introducing the term "ecosystem" in 1935 and is recognized for the important role he played in the professionalism of ecology as an integrated science. Arguing that the model of the organism is inadequate to explain the dynamics of biotic communities, Tansley and others called attention to the diversity of organisms within and among ecosystems, and the need for explanatory models that take into account abiotic factors (e.g., moisture, temperature, light, pressure, wind, chemical cycles), disturbance events (e.g., fire, violent winds, invasive species), and the centrality of energy and biomass flows in ecosystems and their relationships to trophic relationships. Complex models with feedback loops were developed to simulate ecosystem dynamics involving not only relationships among biota but also the role played by abiotic factors and exogenous events.

ALDO LEOPOLD: THE LAND ETHIC

Aldo Leopold wrote near the mid-point of the twentieth century that "There is as yet no ethic dealing with man's relation to land and to the animals and plants which grow upon it." In the final chapter of the *Sand County Almanac* titled "The Land Ethic,"[4] Leopold provided the outline of an ethic centered on man's relationship to the land and its diverse communities of life.

The Land Ethic in Broad Outline

Leopold emphasizes that the "land" needs to be understood as a biotic community to which we belong, not as conqueror, but as a participant and "as a plain member and citizen." When we see land as a community to which we belong, "we may begin to use it with love and respect." Such respect, he emphasizes, extends not only to fellow members but to the community as a whole.

4 Aldo Leopold, *A Sand County Almanac* (New York: Scribner's, 1949). The following quotations from Leopold are taken from pp. 236–248 of this work.

In Leopold, we have the rudiments of an ethic that has both an individualistic and holistic cast. In respecting the whole, we come to respect its many interacting biotic and abiotic parts, including its plants, animals, soils, and waters. "The last word in ignorance is the man who says of an animal or plant: 'What good is it?' If the land mechanism as a whole is good, then every part of it is good, whether we understand it or not. If the biota, in the course of eons, has built something we like but do not understand, then who but a fool would discard seemingly useless parts of it? To keep every cog and wheel is the first precaution of intelligent thinking."

Leopold tells us to "think like a mountain": to realize that misguided attempts to shape nature's ways to isolated human ends can have far-reaching, cascading effects ("tropic cascades"). He emphasizes the importance of ecosystems as wholes and gaining a fuller understanding of the complex relationships and interdependences that exist within (and among) ecosystems. In this context, he recounted how he witnessed the killing of a female gray wolf, with the "fierce green fire dying in her eyes." Reflecting on this experience in the *Land Ethic*, he wrote that "neither the mountain nor the wolf deserved this." In regard to the aggressive campaigns of his day to extirpate wolves, Leopold wrote that "The cowman who cleans his acreage of wolves does not realize that he is taking over the wolf's job of trimming the herd to fit the change. He has not learned to think like a mountain. Hence the dustbowls, the rivers washing the future into the sea."

Leopold expressed concern throughout his career over species extinctions and the biological impoverishment that results from extinctions. He was also critical of the introduction of invasive species that crowd out local and endemic species, land uses that cause soil erosion, water diversions and impoundments that imperil downstream communities of life, and careless use of pesticides and herbicides. Leopold, himself, was a preeminent land manager, having written the first management plan for the Grand Canyon and the Gila National Forest, the first designated legally designated wilderness outside the national park system. Leopold was above all a conservationist who viewed "conservation [as a] state of harmony between man and the land."

The closest that Leopold comes to formulating an ethical principle is his statement that "A thing is right when it tends to preserve the integrity, stability, and beauty of the biotic community. It is wrong when it tends otherwise."

Criticisms and Competing Interpretations

While Leopold's views were, for the most part, initially ignored by the philosophical community, several early commentators wrote unflattering reviews. Australian philosopher H. J. McCloskey wrote, "There is a real problem in attributing a coherent meaning" to his ethic, and English philosopher Robin Attfield called it "dangerous nonsense."

Because Leopold and other ecologists of his day called attention to situations in which human goods were at odds with the goods of larger biotic communities, with some writers drawing the conclusion that the good of the whole always takes precedence over the good of constituent parts, Leopold's ethic was met with wary concern. Opposition was heightened by the use of holistic ideas by some deep ecologists at mid-century to support controversial proposals. A striking example was the publication of a letter published by Earth First! suggesting that AIDS may not be so bad if it proves to slow or reverse exploding population growth. "If the AIDS epidemic didn't exist," Ann Thropy wrote, "radical environmentalists would have to invent one." Radical views such as this led William Aiken to write that Leopold's ethic implies that "massive human diebacks would be a good" and prompted Mark Sagoff to call Leopold's ethic "misanthropic" and Tom Regan to label it as "fascist."

For Leopold's interpreter, Baird Callicott, all of this was unfortunate, noting that "Leopold never intended the land ethic to have either inhumane or antihumanitarian implications or consequences."[5] To set the record straight, Callicott calls attention to statements in Leopold's writings emphasizing that humans are members of many communities, ranging from families to nations to larger biotic communities, with each form of association imposing its own set of moral requirements. "From the biosocial evolutionary analysis of ethics upon which Leopold builds the land ethic," Callicott wrote, "[the land ethic] neither replaces nor overrides previous accretions. Prior moral sensibilities and obligations attendant upon and correlative to prior strata of social involvement remain operative and preemptive."[6] He continues, "The duties correlative to the inner social circles to which we belong [generally] eclipse those correlatives to the rings farther from the heartwood when

5 Baird Callicott, "The Conceptual Foundations of the Land Ethic," in *Foundations of Environmental Philosophy*, Frederick A. Kaufman, ed. (New York: McGraw-Hill, 2003), p. 270.
6 Ibid., p. 278.

conflicts arise. … Family obligations in general come before nationalistic duties and humanitarian obligations in general come before environmental duties."[7]

Writers have questioned whether Callicott is accurately expressing Leopold's position when he says that family obligations generally come before nationalistic and humanitarian obligations, and that the latter generally come before environmental obligations (i.e., whether this view is more Callicott's than Leopold's); however this may be, Leopold's sketch of an ecocentric ethic did not provide a principle or set of principles for resolving conflicts of interest and obligation. Leopold did make it clear, however, that he thought Gifford Pinchot's adoption of the principle of utility as a guiding principle in natural resource management is ill-advised.[8]

Callicott interprets Leopold as holding the view that integrity, stability, and beauty are objective values in ecosystems; that they are inherent values in ecosystems that ought to be respected and protected. Bryan Norton argues that this is a misreading of Leopold. In Norton's words, "While I agree with Callicott's identification of integrity as the key concept in environmental ethics, we nevertheless differ strongly regarding how to interpret this conceptual centerpiece of the land ethic." He continues: "Callicott believes that, by attributing integrity to the biotic community, taken as a whole, Leopold stepped across the line to nonanthropocentrism and declared his moral allegiance to the hypothesis that nature has inherent value."[9] As Norton sees it, Leopold was not abandoning an anthropocentric perspective, but rather was making "a practical remark on the proper focus of conservation."

In regard to Leopold's association of right action with actions that preserve the "integrity, stability, and beauty of the biotic community," critics have argued that *integrity* is too vague a term, that *stability* needs to be scrapped because ecosystems seldom settle into stable homeostasis, and that *beauty* is an intrinsic value that exists only in the eyes of human beholders, not an objective value that inheres in ecosystems. In view of these criticisms, the differing interpretations of Leopold by Callicott and Norton, the question

7 Ibid., p. 278.

8 The views of Gifford Pinchot, President Theodore Roosevelt's secretary of the Interior, are discussed in Section D.1, which follows.

9 Bryan Norton, "Integration or Reduction: Two Approaches to Environmental Values," in *The Environmental Ethics & Policy Book*, Donald VanDeVeer and Christine Pierce (Thomson Wadsworth, 2003), p. 246–47.

addressed next is whether it is defensible to claim that ecosystems are morally considerable in their own right.

ECOSYSTEMS AND MORAL STANDING

The interests and observations of ecologists in the later decades of the twentieth century posed new challenges for the earlier view that ecosystems progress through orderly stages of succession and eventuate in conditions of stable equilibrium. Science historian Donald Worster recounts how research in ecology near the turn of the twenthieth century was leading to an unraveling of the view of ecosystems set forth in the classical ecological studies of the 1960s, '70s, and '80s. In his words, "Ecology is not the same as it was. A rather drastic change has been going on in this science of late—a radical shifting away from the thinking of Eugene Odum's generation, away from its assumptions of order and predictability, a shifting toward what we might call a new *ecology of chaos*."[10] Ecologists are observing a patchwork of chopped-up landscapes, each composed of biota struggling to survive among invasive species and a barrage of external perturbations. Ecologists are focusing on disturbances such as fire, volcanic eruptions, violent wind, invasive species, and the restlessness of climate. Population biologists are seeing lots of individual species, each doing their own thing, but not an emergent collectivity or any strategy to achieve equilibrium and balance. Worster posed several unsettling questions: "What is there to love or preserve in a universe of chaos? ... [Why] not go ahead with all our private ambitions, free of any fear that we may be doing special damage? ... Does the tradition of environmentalism to which Muir belonged, along with so many other nature writers and ecologists of the past—people like Paul Sears, Eugene Odum, Aldo Leopold, and Rachel Carson—make sense any longer?"[11]

In questioning the reasonableness of attributing moral standing to ecosystems, Holmes Rolston acknowledges there are reasons for denying that the attribution of moral is defensible. "Ecosystems do not have experiences and cares as do animals," he says, and "they do not have genomes and organized

10 Donald Worster, "The Ecology of Order and Chaos," in *Earth Ethics: Introductory Readings on Animal Rights and Environmental Ethics*, James P. Sterba, ed. (Prentice Hall, 2000), p. 162.
11 Ibid., p. 167.

centers as do both animals and plants." "In animals the heart, liver, muscles, and brain are tightly integrated, as are the leaves, cambium, and roots in plants. But the so-called ecosystem community is pushing and shoving between rivals, each aggrandizing itself, or else seems to be all indifference and haphazard juxtaposition—nothing to call forth our admiration. … An ecosystem seems to be "too low a level of organization to be the direct focus of moral concern."[12]

"But this is to misunderstand ecosystems," Rolston goes on to say. "It makes the mistake of looking at one level for what is appropriate at another … of looking for characteristics associated with organisms and, not finding them, concluding that what is being studied is devoid of value significance." In Rolston's view, "The evolutionary ecosystem spins a bigger story; limiting each kind, locking it into the welfare of others, promoting new arrivals, increasing kinds and the integration of kinds." "An ecosystem has no head," Rolston continues, "but it heads towards species diversification, support, and richness." Though not a superorganism, it is a kind of vital field … where "order arises spontaneously and systematically as many self-concerned units jostle and seek to fulfill their own programs, each doing its own thing and forced into informed interaction." In ecosystems, one finds "co-action in adapted fit: predator and prey, parasite and host, grazer and grazed, are contending forces in dynamic processes in which the well-being of each is bound up with the other—coordinated as much as heart and liver are coordinated organically."

Developments in complex systems theory provide a scientific framework for reconstructing new meanings for the terms "integrity" and "stability," and for countering the anthropocentric denial of human-independent value. Ecosystems maintain integrity and achieve relative stability as long as their capacity for self-organization is not overcome by exogenous factors, or as long as they do not finally succumb to entropic degradation. Without the kind of ordering latent in self-organization, ecosystems, as all complex organic systems, disintegrate and lapse into chaos. They are indeed subject to disturbance, but as self-organizing entities, they are capable of establishing and maintaining an ordered integrity.

12 Holmes Rolston, "Value in Nature and the Nature of Value," in *Philosophy and the Natural Environment*, Robin Attfield and Andrew Belsey, eds. (Cambridge University Press, 1994), pp. 13–30.

Using the terminology of chaos-complexity theory, what functions as a primary attractor or an ordered end-state toward which the ecosystem as a whole tends is coadaptation for mutual benefit. Mutually advantageous relationships among a multitude of organisms are of value to the system as a whole and to its inhabitants and participants. The value is system-generated; it inheres in the system and is both a product and process whose existence does not depend on human existence. Processes of self-organization in ecosystems can be disrupted, diminished, or destroyed by exogenous forces, and, when this occurs—when the "integrity" of ecosystems disintegrates—their constituent populations are made less well-off.[13]

Viewed in an evolutionary perspective, Rolston characterizes ecosystems as "spontaneously organizing systems of interrelated parts, simultaneously persisting and evolving through changes over decades and centuries" and as "dynamic collection[s] of species, each selected over evolutionary history to do rather well in the niche it inhabits, an adapted fit."[14] Furthermore, "There are ordered regularities (seasons returning, the hydrologic cycle, acorns making oak trees, squirrels feeding on the acorns) mixed with episodic irregularities (droughts, fires, lightning killing an oak, mutations in the acorns) ... and over longer scales there are climate changes, respeciation, new niches generated and occupied."[15]

As self-organizing systems that enable diverse life-forms to coexist, coadapt, and coevolve in relations of mutual advantage , it is concluded that ecosystems are morally considerable in their own right. Mutually advantageous coadaptation is a fundamental, naturally occurring value in ecosystems, and it is a phenomenon worthy of human moral respect. We value institutional arrangements that engender relations of mutual benefit in human life, and

---

13 It was contended in Chapter 4 that Sectionan ecology that relies exclusively on concepts and methodologies derived from the reductionist-mechanistic paradigm is an inadequate, self-limiting science. Ecosystems cannot be fully understood by isolating individual components and studying these in exacting detail from each of several disciplinary perspectives, nor can relationships between individual components be fully understood apart from the role they play in larger relational contexts. An ecology that acknowledges that ecosystems (wholes) have systemic characteristics that are not reducible to components (parts) in isolation is better positioned to understand the complex dynamics of ecosystems and the synergies they generate.

14 Holmes Rolston, "The Land Ethic at the Turn of the Millennium," in *Environmental Ethics: Divergence and Convergence*, Richard G. Boltzer and Susan J. Armstrong (McGraw Hill, 1998), p 395.

15 Ibid., p. 396.

there is no good reason for not acknowledging the importance of this phenomenon in nonhuman communities of life.

DIVERGING VIEWS: PRESERVATION VS. CONSERVATION

As part of decades of discussion and debate associated with landmark environmental legislation,[16] a tension has developed between "preservationists" and "conservationists." The former emphasize that intrinsic values in nature merit protection and preservation; the latter emphasize that nature's resources require protection because they are essential for human survival and well-being. The following discusses how this tension has played out in famous policy debates over the years.

Early Wilderness Debates

An early example was the debate between Gifford Pinchot and John Muir during the first decade of the twentieth century over a proposal to dam the Hetch Hetchy River in a valley adjacent to Yosemite National Park in California. Pinchot was the head of the US Forest Service at the time and the leader of the conservation movement dedicated to wise use of resources for human benefit.[17] In the case of Hetch Hetchy, he argued that damming the river was the most efficient means of providing much-needed water for millions of current and future residents of the San Francisco area and for irrigated agriculture.[18] For Muir, founder of the Sierra Club and leader of the growing preservation movement, damming the Hetch Hetchy would be a terrible mistake. For decades, Muir had written passionately about the majesty and irreplaceability of wilderness areas, about the intrinsic values

16 This legislation included the 1964 Wilderness Act, the 1969 Environmental Protection Act, the 1970 Clean Air Act, the 1972 Clean Water Act, and the 1973 Endangered Species Act.

17 Pinchot believed that natural resources in the public domain should be managed to promote "the greatest good for the greatest number of humans for the longest time." In *Training of a Forester*, Pinchot wrote that

18 "The central idea of the Forester, in handling the forest, is to promote and perpetuate its greatest use to men. His purpose is to make it serve the greatest good of the greatest number for the longest time … [as] was foreseen from the beginning by those who were responsible for inaugurating the conservation movement."

they embodied and their spiritual and aesthetic value for humans. For him, natural resources were not commodities, and their value transcended their instrumental value for humans. Muir's personal relationship with President Theodore Roosevelt played a role in the establishment of Yosemite as a national park.[19] In the case of Hetch Hetchy, however, the utilitarian and anthropocentric views of Pinchot prevailed, and the dam was built.

A similar debate played out several decades later in connection with a bid by Walt Disney Enterprises to develop the Mineral King Valley in the Sierras as a ski resort, a plan that would involve building an access road and high-voltage line across a portion of Sequoia National Park. Soon after the US Forest Service approved the development, the Sierra Club filed suit in a federal court to stop portions of the development that it argued were inconsistent with wilderness preservation, contending that the Forest Service gave disproportionate weight to economic factors and that no weight was given to the well-being of plants and animals and to the intrinsic value of a unique and irreplaceable ecosystem. This time the preservation argument carried the day and the development did not go forward.

Continuing Debate

Subsequent to the Pinchot-Muir era, conservation has generally been associated with efficient resource use for human benefit, while preservation has been linked with initiatives to limit human use and appropriation of natural resources, with the objective of preserving natural values and enabling natural systems to function on their own.

The tension between the two resource-management orientations is further illustrated in an academic exchange between Baird Callicott and Holmes Rolston at the end of the twentieth century. In his article, "The Wilderness

19 Today there are fifty-eight national parks in the US managed by the National Park Service. The first national park, Yellowstone, was created by Congress and President Ulysses S. Grant in 1872, followed in 1890 by Sequoia and Yosemite. In addition, 756 wildernesses have been set aside by congressional acts. These wilderness set-asides are variously managed by the Forest Service, the Bureau of Land Management, and the Park Service in the Department of the Interior and by the Fish and Wildlife Service in the Department of Agriculture under the umbrella of the 1963 Wilderness Act. The Eastern Wilderness Act of 1975 relaxed the definition of "wilderness" to permit the inclusion of areas that could be returned to a "primeval" state through preservation.

Idea Revisited: The Sustainable Development Alternative,"[20] Callicott argued that the concept of wilderness preservation is flawed because it dichotomizes man and nature, overlooks the effect that indigenous peoples had on the lands they occupied, and fails to recognize that ecosystems constantly change over time. To resolve the quarrel between conservationists and preservationists, Callicott recommends that conservation be reinterpreted in terms of the concept of sustainable development—which, he claims, is consistent with Leopold's published and unpublished writings—and that the concept of wilderness preservation, long associated with John Muir, be abandoned.

In a rebuttal article,[21] Rolston addresses Callicott's several criticisms of preservationist thinking. First, recognition of obvious differences between human culture and wild nature does not amount to the radical kind of dichotomization alluded to by Callicott. Second, early humans obviously did modify the environments in which they lived, but not on the scale and with the degree of irreversibility that humans living in industrialized societies modify the environment today. And, third, it is absurd to claim that preservation-minded people do not know that ecosystems change, or that students of natural history somehow object to changes that occur in geological time. Rolston finds merit in the concept of sustainable development, but not in a conception that links sustainability exclusively to economic activity, nor one that insists that nature has no value apart from human valuers, nor one that supposes that humans can improve upon nature. Rolston also takes exception to Callicott's contention that Leopold opposed wilderness preservation, asking why Leopold would have written the following if, indeed, he was anti-preservationist: "I am asserting that those who love the wilderness should not be wholly deprived of it, that while the reduction of wilderness has been a good thing, its extermination would be a very bad one, and that the conservation of wilderness is the most urgent and difficult of all the tasks that confront us."

Essentially the same debate regarding nature's value lingers in the field of conservation biology today, as illustrated by remarks made by Jane Lubchenco in an interview conducted by Yale University's newsletter *Environment 360*[22]

20 Baird Callicott, "The Wilderness Idea Revisited: The Sustainable Development Alternative," in *Reflecting on Nature Readings in Environmental Philosophy*, Lori Gruen and Dale Jamieson, eds. (Oxford, 1994), pp. 252–264.

21 Holmes Rolston, "The Wilderness Idea Reaffirmed," in *Reflecting on Nature Readings in Environmental Philosophy*, Lori Gruen and Dale Jamieson, eds. (Oxford, 1994), pp. 265–278.

22 The full text of the interview is available at http://e360.yale.edu/feature/interview.

concerning a petition signed by some two hundred forty practitioners in conservation biology calling for unification.[23] Lubchenco comments that "Individuals who believe passionately that we should be conserving nature for its own intrinsic value were legitimately concerned that focusing on the value of nature to people not only cheapened the moral and ethical arguments for protecting biodiversity, but also would open the gates to lack of protection if something was not useful to people." "On the other end of the spectrum," she continues, "there's been a huge amount of new work devoted to understanding better the myriad ways in which people benefit from nature and are trying to incorporate that knowledge in work toward healthy ecosystems because it's in our interest to do so. ... Over time, the two perspectives—the first being the focus on intrinsic values, and the second's focus on instrumental values—have gotten more and more polarized."

In concluding her remarks, Lubchenco notes that Peter Kareiva has been a strong champion of the instrumental view. "But," she continues, "he, like me, actually has a pretty inclusive view of this, which is that intrinsic values are, in fact, really important ... but that those values alone have not stemmed the loss of biological diversity and those arguments alone are not sufficient in appealing to the breadth of audiences and stakeholders that are really needed to arrest the loss of both species and functional ecological systems."[24]

Biodiversity Loss and Conservation/Preservation Priorities

Chapter 2, Section D ("Loss of Biodiversity: The Holocene Extinction"), presents an extended discussion of biodiversity loss, institutional responses to the problem, and the debate over conservation priorities. This discussion need not be recounted here, except to say that an ecocentric (ecosystem-oriented) orientation in ethics focuses on substantive moral issues of great importance.

---

23 Lubchenco organized the petition, along with Peter Kareiva, former chief scientist of The Nature Conservancy. The petition was published in the journal *Nature* prior to Yale's interview with Lubchenco.

24 For additional background on the intrinsic vs. instrumental debate, see Keith Kloor's "The Battle for the Soul of Conservation Science" at http://issues.org/31-2/kloor/. Kloor's article contains an extended discussion of the contributions of Michael Soule, a founding father of conservation biology, and Peter Kareiva, chief scientist for The Nature Conservancy, along with an overview of important milestones in the development of conservation biology and a discussion of the prospects of biological diversity and creativity in the Holocene era.

9 CAPITALISM, MARKETS, AND GOVERNMENT

This chapter addresses arguments regarding the respective roles that markets play, or should play, in resource management decision making. These arguments play out in the context of a capitalist society. This chapter begins, accordingly, with a discussion of capitalism.

CAPITALISM AND ITS CRITICS

In portraying the ideal characteristics of a capitalist economy, the following assumptions are made by free market economic theorists:

- *Private ownership*. Productive assets (capital, land, machinery) are privately owned.
- *Profit motive*. Owners are motivated out of self-interest to increase their wealth by producing products that people want and to selling those products and services at a profit.
- *Demand-driven supply*. The supply of products and services are driven by consumer preference and willingness to pay.
- *Robust competition*. Competition among producers creates consumer choice and imposes market discipline, requiring suppliers to focus on cost control, product quality, and consumer satisfaction.
- *Market freedom*. Market entry is not limited by monopolization and buy/sell transactions are free of force and fraud.

- *Minimal government regulation.* Regulation is limited to what is necessary to prevent force, fraud, and monopolization and to keep dangerous products off the market.
- *Rational economic decision making.* Both consumers and producers make informed buy/sell decisions.

In *The Wealth of Society*, Adam Smith posed the question more than two centuries ago why some countries have greater wealth than others and why the citizens living in flourishing economies enjoy high standards of living while citizens of other countries languish in poverty. His answer was that wealthy nations have economies organized along capitalist lines, and that poor nations do not. The description that Smith gave of capitalism—how a capitalist economy can and should work—foreshadowed the modern characterization just summarized.

In Smith's view, a capitalist economy will tend to promote the greatest good for the greatest number of people in society.[1] In a fully functioning capitalist economy, it is as if there is an "invisible hand" operating to promote the general good: individual and societal wealth are created and needs are met. Smith emphasized that humans are motivated not only by self-interest—a positive driver in business—but also by benevolent feeling—a positive force in social life.[2] While wealth will naturally concentrate in a competitive free market economy, Smith maintained, the most successful will, in their public lives, be motivated out of benevolence to share their wealth with the least fortunate members of society in the form of charitable and philanthropic contributions. Such public benevolence is, Smith emphasized, no less a part of the working of the "invisible hand" than is self-interest in business. As later theorists have framed it, the wealth of successful entrepreneurs "trickles down" to all members of society, not only through capital reinvestment to create more jobs, goods, and wealth, but also through the public good works that wealth enables. Everybody is made better off in a capitalist society than under any other system of economic organization.

Critics of capitalism paint a different picture. Capitalism is dressed up to look good in theory, but falters in practice. Large multinational corporations

---

1 Smith was a near contemporary of Jeremy Bentham and was probably familiar with Bentham's formulation of the principle of utility.

2 Smith was also the author of *A Theory of the Moral Sentiments*, in which he developed a theory similar to that developed by his younger contemporary David Hume.

have amassed enormous economic and political power and bear little resemblance to the small-scale businesses of Adam Smith's day. Through their campaign contributions, lobbying, and the "revolving door," today's large corporations exercise extensive self-serving influence in all branches of government. Despite antitrust legislation, they engage in monopolistic practices to eliminate competitors, and their wealth enables them to mobilize public relations campaigns to garner support for corporate policies and to minimize the influence of critics.

Moreover, given inevitable information disparity between sellers and buyers in the technological society in which we live, sellers regularly take advantage of buyers through deceptive advertising and other questionable marketing and sales practices. Also, the effectiveness of supply-side marketing drives demand (not vice versa as postulated in theory), and has created a society awash in nonessential services, the consumption of resources at unsustainable levels, and the generation of enormous quantities of wastes and pollutants.

Corporations operate in a political climate, moreover, that generally enables them to pass the "external costs" of doing business—health, safety, and environmental costs—to other parties: people injured or made sick by products and by-product pollutants, taxpayers who foot the bill for governmental agencies mandated to manage these externalities, and a less productive and resilient natural world.

Still another frequently voiced criticism of contemporary capitalistic society is that it has created a society of "haves and "have-nots": a society in which the wealthiest 1 percent of the population owns 40 percent of total national wealth, while the bottom 40 percent have negative net worth.

RESOURCE MANAGEMENT: THE ROLE OF MARKETS

Gifford Pinchot believed that national forests, the minerals they contain, and the recreational and other uses they afford should be managed in a manner that promoted "the greatest good for the greatest number for the longest time" and emphasized that this goal dictates using forest resources in an economically efficient manner. In carrying out this mandate, Pinchot further believed that managers should rely primarily on the judgment and expertise of professional foresters.

Resource economist Peter O'Toole challenged Pinchot's belief that forest management policy should rest in the hands of forestry professionals. Summarizing his five years of analyzing the activities of the US Forest Service, O'Toole wrote, "I've visited national forests in every part of the country and have seen costly environmental destruction on a grand scale. Money-losing timber sales are reducing scarce recreation opportunities, driving wildlife species toward extinction, and polluting waters and fish habitat."[3] He continued: "My economic research has convinced me that Americans can have all the wilderness, timber, wildlife, fish, and other forest resources they want. Apparent shortages of any of these resources are due solely to the Forest Service's failure to sell them at market prices."

For O'Toole, two major impediments stand in the way of "letting the market decide": (1) the Forest Service is hamstrung by its responsibility to Congress and the competing and sometimes contradictory demands of the diverse interests Congress represents; and (2) forest land management is driven by its budget, much of which comes from revenues obtained from below-cost timber sales. The root problem, O'Toole believes, is that public land management does not function within the context of a free market economy. If market forces were allowed to operate, timber sales would reflect true market value and the users of forest resources for recreation and other uses would pay in the form of market-dictated user fees.

O'Toole cites the writings of resource economists, in particular John Baden and Richard Stroup, who argue it is a mistake to treat natural resources as "public goods" to be managed for the public welfare by experts within bureaucracies. The rational alternative, they maintain, is to privatize resources and rely on private owners to use their property in the ways that best serve their interests, as these interests are expressed in a free, open, and competitive marketplace.[4]

William Baxter also advocates a market-based approach to the solution of environmental problems: "To assert that there is a pollution problem or an environmental problem is to assert, at least implicitly, that one or more

---

3 Randall O'Toole, *Reforming the Forest Service* (Washington, DC: Island Press, 1988), p. 14.

4 For more on the views of Stroup and Baden, see their book, *Natural Resources: Bureaucratic Myths and Environmental Management* (San Francisco: Pacific Institute for Public Policy Research, 1983).

resources are not being used so as to maximize human satisfactions."[5] The challenge, he says, is to find an "optimal level of pollution": "just those amounts that attend a sensibly organized society thoughtfully and knowledgeably pursuing the greatest possible satisfaction for its human members."[6] Baxter emphasizes the need to think in terms of trade-offs, in which the costs of reducing water or air pollution are viewed as what would have to be given up to achieve these objective. The optimal level of pollution is that point at which the next increment of pollution results in a net loss in overall societal satisfaction (the "Pareto optimum"). For Baxter, economic analysis of transactions in a free market economy is the only reliable gauge of the satisfaction of interests: "If individuals in a society are free to engage in whatever exchanges of resources are mutually satisfactory for themselves, then, at least in theory, every resource in society will be deployed in the way that yields the greatest possible human satisfaction."[7]

Cost-benefit analysis is pervasive in both corporate and public decision making, the former conducted from the perspective of self-interest and the latter from a perspective of utilitarianism. The White House Office of Management of Budget conducts cost-benefit analyses of federal agency budget proposals in preparing the administration's budget proposal to Congress, and the Congressional Budget Office similarly scrutinizes legislative proposals proposed by the president and various congressional committees. Critics outside government frequently use cost-beneficial analysis to critique government policy. For example, it was argued that a logging ban to protect the northern spotted owl failed to demonstrate that the benefits of the ban outweigh its costs.

Cost-benefit analysis requires assigning monetary values to all relevant costs and benefits. Markets often dictate the monetary values to be assigned (e.g., the monetary value of a board foot of redwood timber or a ton of borax), but not always. When market values cannot be derived for essentially non-monetary values (e.g., protection of an endangered species or opportunities for hiking and camping in a wilderness area), economists rely on willingness to pay or some other contrived market proxy.

---

5 William F. Baxter, *People or Penguins: The Case for Optimal Pollution* (New York: Columbia University Press, 1974), p. 17.
6 Ibid., p. 8.
7 Ibid., p. 27.

The theoretical and practical objections to exclusive use of cost-benefit analysis are numerous. Mark Sagoff argues that such analysis rests on a serious confusion between wants or preferences, on the one hand, and beliefs and values on the other. When environmentalists argue that a national forest should be preserved for its aesthetic or symbolic meaning, they are stating a conviction about a public good, not merely expressing personal wants. Sagoff emphasizes that people are both citizens and consumers; they have personal preferences as consumers, but they also have social and environmental concerns that are sometimes more important to them than personal desire. Sole reliance on economic analysis as a decision-making tool is, in Sagoff's view, not only a distortion of human nature but also a danger to a democratic way of life. As Sagoff puts it, "Our environmental goals—cleaner air and water, the preservation of the wilderness and wildlife, and the like—are not to be construed simply as personal wants or preferences; they are not interests to be 'priced' by markets or cost-benefit analysis, but are views or beliefs that may find their way, as public value, into legislation. These goals stem from our character as a people, which is not something we choose, as we might choose a necktie or a cigarette, but something we recognize, something we are."[8] As citizens, people are rightfully disturbed by economic decisions that leave no room for discussion and debate.

The plethora of objections that moral philosophers have leveled at cost-benefit analysis is summarized by Ernest Partridge in this way: cost-benefit analysis reduces all value into cash, thus factoring morality out of policy considerations; cost-benefit analysis measures aggregate consumer preferences to the exclusion of community/citizen values; cost-benefit analysis and economic analysis are descriptive, indicating what consumers value (economically) rather than what they ought to value (normatively); and cost-benefit analysis discounts future costs and benefits at the peril of future generations.[9]

RESOURCE MANAGEMENT: THE ROLE OF GOVERNMENT

In a democratic society, government's indispensable role in the formulation and enforcement of resource management policy is that of providing a means

8 Mark Sagoff, *The Economy of the Earth* (New York: Cambridge University Press, 1990), pp. 28–29.
9 See Partridge's "Future Generations" in *A Companion to Environmental Ethics*, Dale Jamieson, ed. (Blackwell, 2001), for a fuller account of these criticisms.

of balancing the oftentimes competing and conflicting beliefs and preferences in society. Granted that resource management policies are, or ought to be, worked out in a democratic political process, the following discussion focuses on the role that government has, in fact, played and theorists argue it should play.

Few, if any, market-oriented environmental theorists advocate a completely unregulated economy, the kind of nineteenth-century laissez-faire capitalism that Pinchot, Muir, and other environmentalists of the time were united in opposing. It is generally agreed that regulation of business activity is required to prevent force, fraud, and monopolization and to protect citizens from practices that pose serious threats to human health and safety. The argument has shifted from "no regulation" to minimal regulation coupled with measures designed to harness market forces to prevent serious harms and make business more environmentally friendly. Much of the discussion focuses on how the economy can be "decoupled" from environmental degradation.

Persuaded by advocates of free market economics, legislation has been passed granting expanded administrative authority to government agencies to implement such market-oriented measures as cap-and-trade systems for air emissions, incentives to encourage investments in renewable energy, and taxes on resource users and polluters as a means of "internalizing" costs that now are passed on to taxpayers and other parties.[10]

Arguments and counterarguments play out, of course, in a complex intergovernmental system and a rapidly expanding global context. Numerous factors influence government's role: powerful industrial and business lobbies that have a vested interest in sustaining a fossil fuels-driven economy and reducing the business costs of regulation; the national security issues associated with dependence on foreign oil; the influence of environmental organizations through lobbying, litigation, education, and other forms of activism; and the concerted efforts of the United Nations and other international organizations to deal with global problems, not the least of which is climate change and biodiversity loss.

10 Part of the rationale for making polluters pay is that such measures have the effect of raising the prices for products and that higher prices will, in turn, have the effect of reducing consumer demand for these products, thereby reducing excessive consumerism and its deleterious effects on the environment.

CAP-AND-TRADE SYSTEMS

In a cap-and-trade system, a government authority establishes a cap on total emissions for a given pollutant and then allocates emission allowances among emitters of that pollutant, in which total allocated allowances add up to the not-to-exceed cap and the allocation to each polluter matches its historical emissions levels. If a polluter emits less than its allowance (e.g., because it has invested in abatement technology as a least-cost compliance option), it is free to sell some of all of its allowance to another party whose least-cost compliance option may be that of buying allowances ("offsets").

Pollution allowances are treated as commodities that can be bought and sold in markets—in effect, putting a price on pollution. Compared with a command-and-control regulation that requires polluters to implement best available technology to comply with uniform ("one size fits all") emissions standards, cap-and-trade creates flexibility for producers and will, in theory, minimize pollution and pollution costs.

Notable cap-and-trade programs include the European Climate Exchange (initially set up under the Kyoto Protocol on Climate Change and reaffirmed by signatories to the 2015 Paris Agreement) and EPA's "acid rain" program established under Title IV of the 1990 Amendment to the Clean Air Act (targeting carbon dioxide and oxides of nitrogen). China is piloting programs in several provinces, with plans to establish a national program.

Innovative cap-and-trade schemes have evolved. Both voluntary and government-regulated exchanges have been established. Many regulated exchanges incorporate planned reductions in annual caps. Some allow the banking of allowances for future use, thus permitting speculators to buy with the hope of selling at a later time at a profit, and environmental organizations to purchase allowances as a strategy for driving up the price of pollution. Some government-regulated exchanges allocate allowances at no cost, while others sell them at auction. California and several other states have implemented both a carbon tax and a cap-and-trade system as part of an integrated strategy to reduce carbon emissions.

KENNETH ARROW: TOWARD ECONOMIC EFFICIENCY

Kenneth Arrow provides a broad macroeconomic argument for greater reliance on market forces for the solution of environmental problems. He argues that making companies pay for the "externalities" of business enterprise—the adverse health, safety, and environmental costs that are not borne by producers but are passed on to others—is a surer and more efficient way of protecting people and the environment than the alternative of relying on taxpayer-supported government to deal with the problems created by these externalities. The basic reason is that private businesses operate in a competitive environment in which market forces impose discipline and cost control, while government agencies operate in a monopolistic environment in which competition is largely absent and cost control is weak. If the external costs of business were borne by business—if product-related health, safety, and environmental costs were internalized in business decisions and reflected in prices—the overall costs to society would be minimized.

10 SUSTAINABILITY, ECOLOGICAL FOOTPRINTS, AND FUTURE GENERATIONS

The concepts of sustainability, ecological footprints, and future generations are closely linked. Some definitions of sustainability are linked to obligations owed to future generations, measures of sustainability are typically framed in terms of the size of ecological footprints, and concern is mounting that ecological footprints exceed the capacity of the planet to sustain the levels of production and consumption that footprints measure.

The imperative of exercising caution in regard to the risks posed by unsustainable development has brought the principle of precaution to the forefront of discussion. This principle—its meaning and implications for practice—is taken up in this chapter.

The chapter's discussion concludes with a discussion of the bioregional movement in which many of the ideas about sustainable living are incorporated.

SUSTAINABLE RESOURCE USE

The term "sustainability" is employed in numerous contexts—ranging from the carrying capacity of the planet to the sustainability of political systems, economies, economic sectors, ecosystems, gardens, and lifestyles, with numerous definitions, standards, and metrics.

In general, a sustainable activity is an activity that can be continued indefinitely: one whose time and resource expenditures makes continuation of that activity and the realization of its intended benefits possible over a long time horizon. Applied to resource utilization (timber harvesting, fishing, groundwater pumping, etc.), sustainability refers to usage rates that permit long-term utilization of the resources in question, thus enabling the activities that are dependent on these resources to continue.

Classical resource economics frames the question of sustainability in terms of growth: how production can keep pace with growing demand on a sustainable basis, given the constraints imposed by competition for scarce natural resources, capital, and labor. Several key assumptions are made in classical resource economics: (1) that the resources needed for unabated economic growth exist in abundance and are replenished and renewed by processes in the natural world; (2) that the natural environment will be able to absorb the waste by-products of economic growth without diminishing regenerative capacity; and (3) that when sporadic natural resource scarcities occur, technology can be counted on to provide usable resource substitutes.

Herman Daly challenges the "classical" view by drawing a distinction between "weak" and "strong" sustainability. Weak sustainability is said by Daly to rest on the assumption that man-made capital can substitute indefinitely for shortages of natural capital, while strong sustainability recognizes that natural capital is a critical limiting factor of sustainable economic activity. A weak sustainability theorist contends, for example, that a lack of fish can be dealt with by building more fishing boats and implementing better fish harvesting methods. A strong sustainability theorist replies that more fishing boats and improved technology are useless if there are too few fish in the ocean, insisting instead that catches must be limited to ensure adequate fish populations for tomorrow's fishermen and species perpetuation.

As noted in Chapter 2, Daly believes that growth becomes uneconomic and eventually impossible when rates of consumption of renewable resources exceed renewal rates, when rates of consumption of nonrenewable resources exceed the rate of development of resource substitutes, and/or when rates of waste generation exceed the capacity of ecosystems to absorb and recycle those wastes.[1] William Rees agrees with Daly. According to Rees, "Our technological prowess feeds the common belief that humans and their economies have become increasingly independent of nature; in actuality, however, modern

1 For a fuller exposition of Daly's views, see his article "Sustainable Growth: An Impossibility Theorem" in *Environmental Ethics: Concepts, Policy, Theory*, Joseph DesJardins, ed. (Mountain View, CA: Mayfield, 1999), pp. 423–427.

high-income economies are becoming increasingly indebted to nature."[2] Rees goes on to explain how technology and trade have merely obscured his relational truth by displacing negative consequences of growth to distant systems and the future. The root problem, Rees says, is that economies are growing at faster rates than the ecosystems in which they are embedded, and, exacerbating the problem, waste discharge and pollution back into the environment impairs ecosystem productivity.

Both Daly and Rees suggest that we think about ecosystems in the way we think about financial investment: when one invests money, it is with the expectation that one's capital (principal) will grow and will yield a continuing stream of income (interest). If we live off the interest generated by natural resources and at the same time diminish the capital that generates the interest, we undermine the possibility of a growing and continuing return on investment. As Rees put it, "Sustainable development is only possible if we are willing to live on the interest of our remaining ecological endowment without depleting this endowment."

Sustainability has come to be seen, more broadly, as a process of maintaining balance in resource use, the direction of investments, technological orientation, and institutional change with the goal of enhancing current and future potential to meet human needs and aspirations. The 2005 World Conference on Social Development referred to "three pillars of sustainability" —social, economic, and environmental—as a conceptual device for emphasizing the need for the recognition of limits and the reconciliation of demands from these overlapping domains. The three pillars of sustainability (the "triple bottom line") have become the basis for numerous sustainability standards and certification systems in recent years, primarily in the food industry

One of the most historically important definitions of sustainability is that given by the Brundtland Commission at the conclusion of the 1987 UN World Summit on the Environment (the "Earth Summit"): "Sustainable development is development that meets the needs of the present without compromising the ability of future generations to meet their own needs."[3]

---

2 William E. Rees, "Consuming the Earth: The Biophysics of Sustainability," in *Earth Ethics: Introductory Readings on Animal Rights and Environmental Ethics*, James P. Sterba, ed. (Upper Saddle River, NJ: Prentice Hall, 2000), pp. 382–386.

3 Some writers see sustainable *development* as an oxymoron or as a concept biased toward economic growth. It is suggested by these writers that the term "development" in the

ECOLOGIAL FOOTPRINTS

Rees coined the term "ecological footprint," defining it as a measure of the extent to which resource demand within a region exceeds the availability of resources within that region, i.e., a measure of the extent to which economic activity within a region depends on imported resources. Rees concluded that the ecological footprint of Vancouver, BC, is twelve times greater than the territory it occupies and is dependent on resources *temporarily* available through a global trading system. As urban populations around the world continue to grow and the world's economies grow even faster, shortages of resources will become acute and a cause of worldwide conflict. Because worldwide demand for resources is estimated already to exceed worldwide resource availability (given current patterns and levels of resource use and technological capacity), temporary surpluses available elsewhere are quickly dwindling and cannot be depended upon indefinitely.

As noted in Chapter 1, the Global Footprint Network maintains a widely used methodology to estimate the ecological footprints of individuals, cities, and nations. Calculation of the footprint involves summing the hectares of cropland, forestland, grazing land, fishing ground, and built-up land to produce the amount of food, fiber, and timber consumed by the individual, city, region or nation in question, and adding to this sum the amount of associated carbon dioxide emitted. This footprint is then compared with a biocapacity measure derived by summarizing the land and aquatic resources available to meet demand, again using the same units of measurement (hectares of cropland, forestland, etc.).

The calculated worldwide ecological footprint is 2.0 hectares person per year, compared with 0.4 for Bangladesh, 1.0 for Pakistan, 6.1 for Ireland, 7.1 for San Francisco, 8.05 for the United States, and 10.6 for the United Arab Emirates. Comparing the ecological footprint of an average American to the global capacity measure, the astonishing conclusion is that 3.8 Earths would be required to sustain the average American lifestyle.

The purpose of such analyses is not to astonish or demoralize, but to call attention that we are on unsustainable paths. The personal lifestyles of people

---

Brundtland definition of sustainability be replaced by "activity" and that reference be made to the future needs of both humans and nonhumans.

in the US and other developed nations cannot be sustained, nor can the economies that support these lifestyles.

While the difference that a single American can make is miniscule compared with the macro-effect of cities and nations as a whole, the message the Global Footprint Network communicates is that individuals can make a difference collectively by eating locally produced organically grown, plant-based foods, driving less in more fuel-efficient cars, and pressuring decision makers to adopt policies and enact legislation that reduce consumption, lessens dependence on fossil fuels, improves efficiency, and minimizes waste generation in all sectors of the economy.

FUTURE GENERATIONS

Thinking sustainably and ethically, the needs of posterity need to be taken into moral account. While few people deny this obvious and seemingly simple moral truth, attempts to clarify what obligations we owe to future generations, and what rights, if any, future generations have against us, raises thorny and sometimes paradoxical questions.

Ernest Partridge argues that people today owe basic duties to future generations—duties implying that future generations have correlative rights against the current generation. Before reviewing Partridge's substantive position on the rights of future generations, it is instructive to summarize his rebuttal of claims that is meaningless and nonsensical to claim that future people have rights. The arguments he addresses are called the repopulation paradox, the time-span argument, the no-claims argument, the non-actuality argument, and the indeterminacy argument. Here, in brief, are the claims and Partridge's counterclaims:[4]

- The **repopulation paradox** claims that any policy designed prevent harm in the remote future will affect future matings and births ("the genetic shuffle") and will, thus, repopulate the future with different individuals. The paradoxical conclusion of this argument is that no

4 For an extended discussion of these arguments and counterarguments, see Partridge's article, "On the Rights of Future Generations," at www.mnforsustain.org/partridge_rights_of_future_generations.htm.

single individual will benefit from a policy designed to prevent future harm. Partridge's reply is that moral principles are binding on persons *in general.* Although a "genetic reshuffle" will result in unknowable particular individuals, we are not relieved of the obligation to hand over favorable life prospects to whomever may be living two, three, or more generations from now.

- The **time-span argument** claims that duties and rights cannot meaningfully be said to hold over long time periods between persons with nonconcurrent lives, i.e., those who are denied reciprocal communication and interaction. But, Partridge replies, events enacted within the lifetime of the present generation may, through "time-lag effects," produce long-term devastation for people born a century or more in the future. Examples of such time-lag effects are underground storage of radioactive materials, global warming, mass species extinctions, and the buildup ozone-thinning hydrofluorocarbons in the upper atmosphere. In such cases, the morally relevant factors are foresight, capacity, and choice, not time.
- **The no-claims argument** maintains that "merely potential" posterity cannot make rights claims and that without such claims there can be no rights. But, Partridge replies, we recognize that individuals who are incapable of claiming their rights may have these rights defended by others acting on their behalf. The rights of infants can be claimed and defended by appointed counsel or by public agencies; the rights of animals can be legally represented by organizations such as ASPCA; a person can stipulate in a will that certain funds shall be held in trust for designated purposes (e.g., the education of yet-unborn grandchildren); and the National Park Act of 1916 specifies that the National Park Service shall protect and keep the land in its charge "unimpaired for the enjoyment of future generations."
- The **non-actuality argument** is the contention that since posterity does not exist now, it makes no sense to speak of posterity as having rights now. If the non-actuality argument is interpreted as the charge that posterity is "merely imaginary," the obvious reply that it is virtually certain (not imaginary) that people will exist three or four generations from now. If the non-actuality argument is interpreted as the contention that posterity's rights apply only in posterity's own time, the

rejoinder is that rights and duties apply to *possible* circumstances, not only to "here and now" actual circumstances.

- The **indeterminacy argument** is the contention that rights cannot be ascribed to a class of people who have no identifiable members. Partridge's reply is that the duty not to be negligent or not to harm refers to *anyone* who might be harmed. If I have a duty to clean up a campsite—to leave it at least as clean as I found it—the obligation is not owed to anyone in particular. The morally operative condition here is not the time of the next use, nor the identity of the next camper. The morally relevant factor is that the campsite might be used by an unknown and indeterminate individual with an interest in having a clean campsite.

The questions about duties owed to future generations have been addressed by theorists from a number of ethical perspectives, including utilitarian, communitarian, rights-based, and social contract perspectives. Each of these approaches are briefly reviewed next.[5]

The **utilitarian approach** emphasizes that morally right choices are choices that maximize well-being when the well-being of all potentially affected parties is taken into consideration, future persons included. Critics find the utilitarian approach to be problematic for a number of reasons. First, if equal distributive shares of needed resources are insisted upon, the enormously large number of expected future people leaves virtually nothing for current generations. Second, it is claimed there is no practical basis for taking into consideration the well-being of future individuals whose tastes, preferences, and needs we cannot know with any reasonable degree of certainty. And, third, utilitarianism is said to flounder on the question whether average or total utility should be maximized. Either response leads to what Derek Parfit called "repugnant conclusions." If average utility is to be maximized, Parfit hypothetically surmises, Adam and Eve should not have propagated; they lived in a "better" world than do thousands and millions of descendants who, though they may be quite happy on average, are less happy than were their original ancestors. If total utility is to be maximized, fertile couples

5 An extended discussion of various approaches, including criticisms that have been leveled at each, is discussed by Partridge in another article, "Future Generations," available at gadfly.igc.org/papers/futgens.htm.

living in an already crowded resource-constrained world would be obligated to produce more children so that total happiness can be maximized.[6]

The **communitarian approach** emphasizes that one generation is bound to past and future generations by virtue not only through shared cultural beliefs and institutions, but also by universally shared moral sentiments such as benevolence. Moreover, human experience has a temporal dimension that extends beyond the span of a lifetime, both past and future.[7] Critics reply that institutional and moral frames of reference change over long expanses of time and that it is questionable whether temporally remote generations will share the cultural beliefs and practices of the current generation, particularly in view of the cultural diversity both within and among cultures in today's world.

The **moral rights approach** contends that human beings *qua* human beings are entitled to certain kinds of treatment. Critics have replied that this claim is not plausible if positive rights (rights to be benefited) are referred to, but still questionable if the claim involves negative rights (rights not to be harmed) since we cannot know with sufficient certainty and specificity what will be harmful to future people, or harmful enough, to justify binding constraints on what people may permissibly do today.

The **contractarian approach** emphasizes that we are bound to abide by rules that a rational person, choosing rationally, would give one's intellectual assent to. This approach is illustrated by conclusions reached by John Rawls in his influential book, *A Theory of Justice*. Rawls asks what institutional ordering principles would be chosen if people in the "original position" made a self-interested maximin choice of those principles without knowing what their position would be in the society whose institutions are to be ordered by the principles they choose. He argued that, if people rationally chose this way, they would unanimously insist on several ordering principles: that all are to have an equal right to essentially equal freedoms, that those most disadvantaged by "birth circumstances" are to be made better off or at least not worse off, and all are to have an equal opportunity to pursue their interests and aspirations under conditions of equal opportunity. As Rawls sets up the argument, the choice of ordering principles involves choosing

6 For more on the average vs. total utility problem, see M. Warren's "Do Potential People have Rights?" in the *Canadian Journal of Philosophy*.

7 For a developed presentation of the communitarian perspective, see Avner de-Shalit's *Why Posterity Matters* (London: Routledge, 1995).

behind a "veil of ignorance," which, he explains, requires a sincere attempt to imagine that one does not know what one's gender, race, native intelligence, and other involuntary characteristics will be when they take their place in the society whose institutions (educational, economic, and political) are to be ordered according to the ordering principles they choose.[8] While Rawls did not explicitly include time of birth—whether one is born into Generation X or Generation Y—in the veil of ignorance, some interpreters argue that if this imagined unknown is included in the veil of ignorance, the same rules of fairness (ordering principles) apply to future generations and to people living today. In another context, Rawls suggests that decision makers should regard themselves as if they were "heads of families" with the responsibility for carrying out "rules of just savings." Among such rules, Rawls includes preservation of the gains of culture and civilization, maintenance of just institutions, and dedication of a suitable amount of real capital accumulation to support investments in future needs. Partridge observes that there is no direct mention of the conservation of resources in this list of rules of just savings, though he suggests that conservation is arguably implicit.

Regardless of the theoretical framework employed to clarify obligations owed to future generations, some writers (e.g., Norman Care) raise a motivation problem. Arguing that we can have no psychological bonds of love or concern for indefinite future persons, or no sense of common humanity with persons of an abstract future, Care thinks that motivation to support policies demanding serious sacrifice for the sake of future generations will be lacking. This line of argument is challenged by writers who emphasize that moral concern is rooted in feelings of benevolence and sympathy, and that such feelings extend to future persons. Rawls addresses the question of motivation in his suggestion that decision makers should choose as if they were "heads of families." Critics have replied that parental love extends to children and grandchildren but diminishes to insignificance within a few generations. Partridge himself contends that "self-transcendent love" is not only possible but represents a healthy and normal process of maturation and socialization—that the social origins of a concept of self, the objectification of values in valued objects, and awareness of one's mortality all lead one to

8 Rawls argues that choosing this way is facilitated by the antipathy one would feel about having their future well-being dependent on involuntary characteristics they did not choose and over which they have no control.

believe, or at least hope, that valued states of affairs will flourish beyond the span of one's lifetime.[9]

While there is much that we cannot know about future people, we do know, Partridge says, that they will be humans with well-known biological requirements, that they will be sentient and have a capacity to reason and choose among alternative futures, that they will require functioning ecosystems if they are to flourish, and that they will need the knowledge and skills required to deal with the problems that may occur during their lifetimes.

Given at least this level of understanding of future generations, Partridge states that we have the following duties to future generations, and that future persons have a right to expect that we take these duties seriously. The first is the common-sense principle of nonmaleficence: the principle that we ought not inflict avoidable harm, including environmental and ecological degradation that we know are harmful to us and have no reason to believe will not be harmful to people in the future. Second, we should not jeopardize the opportunity of future generations to meet basic needs. We know they will need energy and food and we should not deplete or degrade natural resources to an extent that the ability of future people to meet basic needs is seriously compromised. Third, we owe to future generations the obligation not to seriously limit their options to address the problems they may face. This obligation entails our continuing investment in scientific and technical education and research. Fourth, we should take actions to minimize problems that we have good reason to believe they will inherit (e.g., problems associated with global warming, radioactive materials, and biodiversity loss) and should otherwise forgo activities known to cause future harm. Fifth, we should favor policies that work to the advantage of both present and future generations and that are the least burdensome to the present generation. Sixth, since none of this will be accomplished unless people today are inclined to "do the right thing," adequate investments should be made in environmental and moral education, with emphasis on critical thinking, natural history, human history, basic science, and respect for free institutions. Sixth, we ought to institute "custodians" or "guardians" of future generations

9 Partridge's views on the ability of present persons to be motivated to act morally toward future generations is developed in his "Why Care about the Future?" in *Responsibilities to Future Generations*, E. Partridge, ed. (Buffalo: Prometheus), 1981.

in the legal system with authority to speak, advocate, and adjudicate on their behalf in courts of law.[10]

THE PRINCIPLE OF PRECAUTION

Precaution has been informally defined as "due caution," "caution in advance", "caution practiced in the context of uncertainty", and "informed prudence." The core idea is that potential harms associated with proposed undertakings should be taken into account by decision makers before proceeding and, if there is significant uncertainty concerning the severity and probability of risk, action ought to be delayed or postponed until it can be judged with confidence that the undertaking is acceptably safe. A related idea is that mitigation plans ought to be in place before "going ahead" in order to deal with unanticipated and/or incompletely understood harms.

A distinction between "strong" and "weak" precaution is made, in which the former places the burden of proof on the party proposing an undertaking (the burden of proving it is acceptably safe) and the latter puts the burden on opponents of the undertaking (the burden of proving it is unacceptably unsafe). The strong version typically *requires* precautionary measures and sometimes establishes liability for environmental harm (in effect, making polluters pay). On the weak interpretation, it is contended that the opportunity costs of not proceeding needs to be given full consideration in cost-benefit analysis and that the practice of discounting future costs and benefits is warranted.[11]

A distinction is also made between precaution as an approach and precaution as a principle. The former views precaution as a frame of reference or as an important consideration that ought to be taken into consideration by decision makers, whereas the latter views precaution as a decision-making

10 It has been noted that Christopher Stone is an advocate for the recognition of legal standing for future generations and for the appointment of guardian-attorneys to represents the interests of future generations in courts of law. It is noted that a state of Washington court ruled that the state has a constitutional obligation and public trust duty to preserve protect, and enhance air quality for current and future generations. For more on this ruling, see http://www.ourchildrenstrust.org.

11 Discounting costs and benefits over a long time horizon (e.g., over two or three generations) is claimed by Partridge and other theorists to be unfairly prejudicial to future generations.

principle that is binding on decision makers. It is noted in this regard that the 1992 Rio Declaration (issued at the conclusion of the United Nation's "Earth Summit") stated that "In order to protect the environment, the precautionary *approach* [italics added] shall be widely applied by States according to their capabilities. Where there are threats of serious or irreversible damage, lack of full scientific certainty shall not be used as a reason for postponing cost-effective measures to prevent environmental degradation."

The 1982 UN World Charter for Nature gave the first international recognition of the principle of precaution in this statement: "When potential adverse effects are not fully understood, the activities should not proceed." Here precaution is treated as a mandated directive for decision makers. The 1998 Wingspread Statement on the Precautionary Principle (convened by the UN Science and Environmental Health Network) and the 2000 European Commission Report on the Precautionary Principle similarly treat precaution as policy directive to decision makers. According to the 1998 Wingspread statement: "When an activity raises threats of harm to human health or the environment, precautionary measures should be taken even if some cause and effect relationships are not fully established scientifically." The 2000 European Union Communication stated: "The precautionary principle applies where scientific evidence is insufficient, inconclusive, or uncertain, or where preliminary scientific evaluation indicates that there are reasonable grounds for concern that potentially dangerous effects may be inconsistent with the high level of protection chosen by the EU."

BIOREGIONALSIM

Bioregionalism is a movement within the broader environmental movement that has attracted growing attention in recent decades. Similar to other movements, it is an admixture of theory and practice with a history involving many players. Various labels are applied, including "region-based sustainable communities," "place-based cultural transformation," and "eco-communalism."

Peter Berg and Raymond Dasmann characterized bioregionalism as "living-in-place" and as "reinhabitation," in which the latter depends upon an understanding of "activities and evolving social behavior that will enrich the life of place, restore its life-supporting systems, and establish an ecologically and socially sustainable pattern of existence within it."[12] Jim Dodge characterized bioregionalism as *resistance* ("against the continuing destruction of wild ecosystems" and "the ruthless homogeneity of national culture") and as *renewal* ("achieved thorough knowledge of how natural systems work, delicate perceptions of specific sites, the development of appropriate techniques, and hard physical work of the kind that puts you to bed at night").[13] In *Dwellers in the Land: The Bioregional Vision,* Kirkpatrick Sale characterized bioregionalism as decentralized government "committed to promoting autonomy, diversity, and the integration of urban, rural, and wild environments on a regional scale."

Bioregionalism sees the bioregion as the scale of decentralization best suited to support decentralized governance aimed at achieving a more just distribution of economic and social goods. And it embraces democratic governance mechanisms that are open to participatory decision making and held accountable to standards of sustainability and self-reliance. These core themes are elaborated in various theoretical frameworks and activist platforms. Ecofeminist writings are sometimes assimilated, as are vestiges of thinking from the earlier deep ecology movement and the views of Murray Bookchin, who emphasizes that environmental and ecological problems are rooted in conceptions of hierarchy that need to be rooted out. Philosophers such as Mark Sagoff, Bryan Norton, and Ernest Partridge lend support to planning and management practices that are sensitive to democratic processes, local values, sustainability principles, and obligations owed to future generations.

Doug Aberley discusses numerous initiatives that have been undertaken to put bioregional ideas into practice.[14] The Planet Drum Foundation began publication in 1979 of its networking periodical *Raise the Stakes.* The Ozark Area Community Congress has been held every year since 1980 to provide a template for locally oriented and place-based organizing. Other

12 Peter Berg and Raymond Dasmann, "Reinhabiting California," in *The Ecologist* 7(10), pp. 339–40.

13 Jim Dodge, "Living by Life: Some Bioregional Theory and Practice," in *CoEvolution Quarterly* 32, pp. 6–12. For Dodge, "Theories, ideas, notions ... have generative and reclaimative values, and certainly a loveliness, but without the palpable intelligence of practice they remain in the nether regions of nifty entertainment or degrade into flamboyant fads and diversions. ... If theory establishes the game, practice is the gamble."

14 See Doug Aberley's "Interpreting Bioregionalism: A Story from Many Voices," available at http://janucc.nau.edu/.

bioregion-based groups have sprung up across the country, each with its own newsletters and "how to" publications. Alberley also comments on how similar initiatives are occurring throughout the world, including Sweden, New Zealand, Australia, and Germany.

"To those who hear only a part of the bioregional story," Aberley says, "or who attempt to analyze bioregionalism only through the filters of academic specialties, it may seem to suffer a host of apparent weaknesses, contradictions, or unresolved conflicts." "For those who take the time to listen to more of the voices," he continues, "or better yet participate in the bioregional movement itself, chaos transforms into something that is properly perceived as a persistent and organic growth of purpose."

Obviously, there are powerful forces that work against bioregionalism, but the movement has staying power and will not go away quietly.

11 ENVIRONMENAL IMPACT ASSESSMENT: A MODELING APPROACH

In contrast with evaluation methodologies that employ a single objective function (e.g., profit maximization) to assess the feasibility of policies and projects, this chapter outlines a modeling approach that incorporates multiple evaluation criteria to evaluate how policies and large-scale projects affect both humans and nonhumans, both negatively and positively. After describing the model's conceptual structure—its input requirements and computed policy-relevant outputs—the discussion turns to how the model can be used in a variety of contexts, including classroom use involving cases. Suggestions are made for developing required model inputs, using the model for value weighting, and avoiding conceptual pitfalls. A detailed case for possible classroom use (titled "Drilling in the Alaskan Native Wildlife Refuge") is provided in the appendix.

MODEL OVERVIEW

The model's conceptual structure involves or incorporates the following:

- Designated biotic entity types potentially impacted by a policy or project (e.g., humans, vertebrate animals, invertebrate animals, plant life, and imperiled species and/or ecosystems);
- Predicted harmful and beneficial (positive and negative) impacts of the policy or project assessed;
- Six designated harm types (physical damage, resource deprivation, functional impairment, physical pain, emotional distress, and life termination) and six designated benefit types (stated as complements or opposites of the six harm types) in terms of which predicted impacts are characterized as harmful or beneficial; and

- Six impact parameters (extent, onset, duration, proliferation, controllability, and remediability) in terms of which the magnitude and probabilities of predicted impacts are scaled (as model inputs).

Table 11-1 exhibits the model's matrix organization. Table 11-2 identifies the harm and benefit types incorporated in the model. Table 11-3 illustrates the scalars used to estimate degree-of-magnitude impacts.[1]

Given probability and parameter estimates for each predicted impact, the model computes a number of policy-relevant indices used to assess the permissibility of a proposed policy or project (P). Four of these indices are shown in Table 11-1: two at the population level and two at the aggregate entity level. At the population level, the scaled harm parameters associated with each predicted impact are summed and this sum is multiplied by the scaled probability estimate for each. At the entity level, the degree-of-magnitude scalars associated with each predicted impact are summed. At the project (or policy) level, two additional indices are calculated: a total harm index calculated by summing the entity-level harm totals and a total benefit index calculated by summing the entity-level benefit totals. Calculation of a benefit/cost ratio is then calculated by dividing total project-level benefits by total project-level harms. If the ratio is not greater than 1.0, the policy or project (P) is judged not to morally permissible (i.e., impermissible). If the ratio is greater than 1.0, the threshold test of permissibility is met (total benefits exceed total harms), leaving open the question of whether other conditions of moral permissibility are satisfied.

MODEL APPLICATION

Before discussing specific guidelines for applying the model and avoiding conceptual pitfalls, a few suggestions of how the model might be used in a classroom environment using a selected case will provide initial clarification.

Classroom Application

In a classroom environment, a multi-round process of inquiry could be organized to assess a proposal to implement a policy or large-scale project.

1 Scaling for use in probability estimation is discussed subsequently.

Table 11-1 Model Illustration

PREDICTED IMPACTS	REQUIRED MODEL INPUT VALUES							CALCULATED INDICES			
(listed under a relevant harm or benefit type)	Scaled Probability Estimates	Scaled Parameter Degree of Harm & Benefit Estimates						Population-Level		Entity-Level	
		Extent	Onset	Duration	Prolif-eration	Control-lability	Remed-iability	Total Harm	Total Benefit	Total Harm	Total Benefit
<u>**Harms**</u>											
Physical/structural damage **Predicted impact:** ⋮											
Resource deprivation **Predicted impact:** ⋮											
Functional Impairment **Predicted impact:** ⋮											
Induced pain **Predicted impact:** ⋮											
Induced distress **Predicted impact:** ⋮											
Life termination **Predicted impact:** ⋮											

(*Continued*)

PREDICTED IMPACTS	REQUIRED MODEL INPUT VALUES							CALCULATED INDICES			
(listed under a relevant harm or benefit type)	Scaled Probability Estimates	Scaled Parameter Degree of Harm & Benefit Estimates						Population-Level		Entity-Level	
		Extent	Onset	Duration	Prolif-eration	Control-lability	Remed-iability	Total Harm	Total Benefit	Total Harm	Total Benefit
Benefits											
Physical repair/recovery **Predicted impact:** ⋮											
Resource access/availability **Predicted impact:** ⋮											
Functional recovery/enhancement **Predicted impact:** ⋮											
Pain relief **Predicted impact:** ⋮											
Stress relief **Predicted impact:** ⋮											
Life continuation **Predicted impact:** ⋮											

Table 11-2 Characterization Of Harm And Benefit Types

HARM TYPES
Physical damage: damage to structural wholes and parts; crushing, tearing, dismembering, burning, freezing; exposure to poisons, toxins, corrosive substances
Resource deprivation: diminished availability of, or access to, essential resources (nutrients, food, water, shelter); interference with conditions necessary for basic needs fulfillment; diminution of naturally occurring processes in ecosystems; habitat destruction
Functional impairment: disrupted growth and development; interference with naturally occurring internal regulatory processes; impeded freedom of movement and expression
Induced pain: feeling pain or intense discomfort: subjection to, or exposure to, conditions and actions directly causing immediate pain

BENEFIT TYPES
Physical repair/recovery: conditions and actions that prevent or abate physical and structural damage; conditions and actions that repair damage and restore structural integrity
Resource availability: conditions and actions that prevent resource deprivation; conditions and actions that restore access to essential resources
Functional recovery: conditions and actions that prevent, stop, reverse, or ameliorate functional impairment
Pain relief: conditions and actions that prevent or relieve pain; conditions and actions that make pleasurable activity possible
Distress relief: conditions and actions that prevent or reduce emotional distress; conditions and actions that contribute to safety, security, and emotional well-being
Life continuation: conditions and actions preventing immediate death, extinction, and ecosystem collapse
Induced distress: suffering, frustration, anxiety, loneliness, etc.; exposure to, or subjection to, conditions causing immediate emotional distress
Life termination: end of life of individual living beings (death); species extinction; ecosystem collapse and dissolution, often leading to radically changed environmental conditions and relationships

Table 11-3 Parameter Degree-Of-Magnitude Scalars

Extent 1 = not very widespread 2 = moderately widespread 3 = very widespread	**The question:** How widespread is the predicted condition (harmful or beneficial) expected to be within the identified population?
Onset 1 = soon 2 = within a year or so 3 = years in the future	**The question:** How soon is the predicted condition expected to occur?
Duration 1 = a short time 2 = a moderately long time 3 = a very long time	**The question:** How long is the predicted condition expected to last if not controlled?
Proliferation 1 = no follow-on harms/benefits expected 2 = some follow-on harms/benefits expected 3 = many follow-on harms expected	**The question:** Is the predicted condition expected to lead (causally contribute) to other harmful/beneficial conditions within the identified population?
Controllability* 1 = prospect of termination/abatement good 2 = prospect of termination/abatement "iffy" 3 = prospect of termination/abatement poor	**The question:** Can the predicted condition be terminated (arrested) or abated (diminished)?
Ameliorability 1 = prospect of remedy/enhancement good 2 = prospect of remedy/enhancement "iffy" 3 = prospect of remedy/enhancement poor	**The question:** Can the predicted condition be remedied (if harmful) or enhanced (if beneficial)?

* Controllability generally applies only to harmful conditions, not beneficial conditions.

The approach might involve dividing a class into an assessment management team and several impact assessment teams, each corresponding to one of aggregate entity types included in the assessment. If, for example, a class has thirty students and five aggregate entity types are included in the assessment, five students would be assigned to each impact assessment team, leaving five students for the management team.

Using questionnaires and forms prepared by the assessment management team in a multi-round process to assemble predicted impacts and develop estimates of probabilities and impact magnitudes, a first round would focus on the assemblage of predicted impacts, and a second round on refining first-round predictions, adding new predictions, and making initial estimates of impact probabilities and magnitudes by entering scalers in the model's cell matrix. A third round would refine second-round model inputs (probability and degree-of-magnitude estimates) based on second-round feedback.

A fourth round might be added, depending on available time in light of other course work and objectives. The management team would play a critical role in compiling and organizing the model inputs (scaled degree-of magnitude and probability estimates) developed by each impact assessment team. It would also be the responsibility of the management team to perform "what if" analyses by making appropriate changes in model inputs and organizing materials in ways that stimulate and encourage questions and critique in class discussion. The overall process is thought to be a fruitful way of eliciting informed judgments, facilitating interaction, and providing opportunities for the class as a whole to question predictions, model inputs, "what if" scenarios, and the significance and implications of the various computed indices, including the overall benefit/harm ratio for the policy or project assessed.

Entity Types and Impact Prediction

Taking practical and logistical considerations into consideration, the aggregate entity types might be limited to humans, vertebrate animals, invertebrate animals, plant life, and locally relevant imperiled species and ecosystems. While a vast host of microorganisms would be excluded by this typology, the inclusion of locally imperiled species and ecosystems would presumably extend to microorganisms that play critical roles related to species preservation or ecosystem health.

Regardless of what biotic entity types are included, the first step is the assembly of predictions of how the policy or project (P) is expected to impact populations subsumed under the several general entity types. For each predicted impact, this requires identifying a condition expected to be brought about by P, specifying what population or populations would be impacted, and characterizing the nature of the harm or benefit. To the fullest extent feasible, predicted impacts should be based on information obtained from credible sources.

In identifying and compiling predicted impacts, attention should be given to keeping predictions at roughly the same level of generality or specificity, particularly when a predicted effect (changed condition) pertains to more than one entity type or to multiple sub-groups within an entity type. If a project or policy were judged to result in starvation for identified populations of both vertebrate and invertebrate animals, for example, it would be methodologically problematic to include as a single prediction starvation for a general class of vertebrate animals, while including separate predictions of starvation for each of numerous particular (less general) classes of invertebrate animals.

Characterization of Impacts as Harmful or Beneficial

A predicted impact requires a description of a changed condition brought about by a policy or project (P) together with an explanation of how the change is thought to be harmful or beneficial to a species-population. An increase of airborne asbestos fibers is an "impact," but the characterization is not complete. A more complete characterization would involve characterizing how this condition is harmful—as leading, for example, to breathing problems and possibly to cancer. While this example is framed in terms of harm to humans, the general point is that predicted impacts require specification of how an identified condition is expected to be harmful or beneficial to an identified population of individuals.

While it is presumed that the harm and benefit categories (characterized in Table 11-2) encompass the major ways a biotic entity can be harmed or benefited, it is not assumed that all biotic entities can be harmed or benefited in all of these ways,[2] and it not assumed that they can be harmed or benefited in only one way. On the contrary, multiple harms or benefits are often involved, sometimes involving complex causal relationships. Clear-cutting to harvest redwoods, removal of mountaintops to expose coal seams, and the draining of wetlands to build homes or commercial centers are examples of large-scale projects that result in physical injury, resource deprivation, functional impairment, and mass death to multiple species populations. When a predicted condition is expected to involve several types of harm or benefit, each should be identified and subsumed under the relevant harm or benefit types.

The generality of the harm and benefit types is advantageous in most harm assessments. The general categories can, of course, be disaggregated. For example, resource deprivation might be broken down by resource type (e.g., diminished access to water, nutrients, shelter). The appropriateness of disaggregation depends on various factors, including an initial scoping of what biotic entities are expected to be harmed or benefited, what the information needs of policy makers are, and how potentially large amounts of information can most effectively be managed.

2 When it is obvious that a harm or benefit type is not applicable to an aggregate entity, the corresponding cell in the impact matrix should be "shaded out" in advance.

Use of Parameters to Measure Degree-of-Magnitude Impact

The model incorporates six parameters in terms of which impact magnitude is measured (scaled) in the model. It is presumed that the identified parameters (extent, onset, duration, proliferation, controllability, and remediability) encompass all of the important factors contributing to judgments about harm magnitude—about *how harmful* an identified harm may be. As Table 11-3 illustrates, the various parameters are associated with distinctly different questions, although an informed answer to one question (e.g., whether the harm is expected to be long-lasting) may presuppose an answer to another question (e.g., whether the harm in question can be eliminated or mitigated).

As with harm and benefit types, each of the parameters might be broken down into more specific categories. For example, controllability might be divided into arrestability and abatability, but such disaggregation would be useful only if special policy questions are pertinent. Certain parameters might also be combined. For example, controllability could be combined with remediability, or onset with duration. While this may be a way of simplifying an assessment, a rigorous assessment would still involve asking the distinctly different questions that parameter estimation involves. Again, the feasibility of disaggregating or aggregating the impact parameters depends on circumstances associated with the policy or project proposal evaluated, the information needs of decision makers or other interested parties, and the logistics of dealing with the complexities involved.

Probability Estimation

As input values to the model, both probability estimates and parameter degree-of-magnitude estimates are needed. Measurement in each case involves the use of scalars. In regard to the task of making probability estimates associated with predicted impacts, a three-point scale is suggested, where a model input value is 1, 2, or 3, where 1 = probability less than 30%, 2 = probability in the 30–70% range, and 3 = probability greater than 70%. A five-point scale could be employed if more exacting probability estimates were thought to be appropriate in light of data availability and the information needs of decision makers.

A three-point scale is also suggested for estimating degree-of-magnitude estimates associated with each impact parameter (extent, duration, etc.). Whatever scale is used, it is essential that it be used consistently and that scalars are clearly defined. The scaled parameter estimates are entered in the appropriate cell of the matrix, i.e., the cell at which the harm type associated with a predict impact intersects with the parameter in question.

Value Weighting

Harm *seriousness* has two dimensions: the magnitude of the harm involved, and how that harm is judged in value terms. In the first sense, "seriousness" is defined by the scalar values that collectively define its magnitude. Harm that is widespread, imminent, and long-lasting, and that is causally related to other harms and not controllable or remediable, is worse than harm that affects a few, is not imminent, is short-lived, does not lead to other harms, and can be controlled and remedied. "Worse" is used here in a non-value sense. To say that something is worse in a value sense is to say that something of value is at stake. The clear-cutting of large tracts of virgin redwoods in an old growth forest in Northern California is troubling to many people because more than jobs, tax revenues, and forest products are at stake. Also at stake is the loss not only of an irreplaceable ecosystem valued for its beauty but also the loss of nonhuman value: the inherent value of the ecosystem itself and the well-being of its many non-human inhabitants. These are value judgments. Harm qua harm is a disvalue, but some harms in some contexts are legitimately judged to be more serious.

Before getting into a discussion of how the model can be used for value weighting, it is instructive to consider reasons for not assigning value weights at all. *First*, the scaling of parameters to gauge impact magnitude, and the use of this information to generate aggregate harm and benefit indices, already summarizes a great deal of policy-relevant information. *Second*, the view held by many that human interests should carry more weight in deliberation than nonhuman interests is accommodated by a methodology that recognizes that humans can be harmed in more ways, and sometimes to a greater degree, than nonhumans. *Third*, value weighting invariably invites the question of whose weights should be used. Should the weights utilized in an assessment be those judged to be appropriate by a coalition of environmental organizations opposing development, or by industries advocating development. And, fourth, it must be acknowledged that public sector decision makers in a democratic society pay attention to voices expressed in public forums, not analytical models alone and not those that incorporate a particular set of value weights.

If the model is used for value weighting, the weights should be assigned to predicted impacts, not to harm types or aggregated entity types. To make value weighting manageable, it is suggested that those impacts having the highest calculated magnitudes be selected for value weighting: for instance, one set of five for negative impacts and one set of five for positive impacts. With pared-down lists of harmful and beneficial impacts in hand, the next step is to assign value weights by assigning a negative value scaler (–1, –2, or –3) to each predicted harmful impact and a positive value scalar (1, 2, or 3) to each predicted beneficial impact, where, on a three-point scale, the assignment of –1 or 1 reflects the judgment that the value significance of the predicted impact is low, –2 or 2 reflects the judgment that the value significance is moderate, and –3 or 3 reflects the judgment that the value significance is high. The assigned scalars are then added to the computed population-level harm and benefit indices to yield value-weighted indices for each predicted impact. Calculation of entity- and project-level indices would, *ipso facto*, be value weighted, including the project-level benefit/cost ratio.

An approach that assigns value weights to entity or harm types runs into problems. If harm types are weighted, all impacts subsumed under a given harm type are given the same weight. If aggregate entity types are weighted, the door is open to unacceptable interspecies bias. The principle of equal consideration—requiring that comparable harms (or benefits) be given comparable weight in moral deliberation—applies to value weighting in the present context. If a situation arises in which a predicted condition is judged to be harmful (damaging, resource depriving, functionally impairing, physically painful, emotional disturbing, of life-ending) for both an identified human population and an identified population of vertebrate animals, the same value weight ought to be assigned.

It is acknowledged that weighting on the basis of a subset of predicted impacts having the highest harm magnitude indices does not eliminate the problem of interspecies bias, but it could be expected to minimize such bias, particularly if alternative sets of weightings provided by diverse interest groups are used to determine whether a proposed policy or project previously (tentatively) judged to be morally permissible continues to pass the moral permissibility test. Weighting along the lines suggested is a useful means of answering this question: Would a proposed policy or project judged to be morally permissible without value weights also turn out to be morally permissible under each of several sets of value weights, or an average of these? A proposed policy or project whose permissibility stands up against a variety of weighting schemes would be one whose moral justifiability is strong (assuming that other permissibility criteria are met—e.g., that no moral rights are violated).

The proposed methodology steers clear of questions of whether the good of a species or ecosystem takes precedence over the good of the individuals making up that species or ecosystem, or whether the good of individuals takes precedence over the collective whole. Value distinctions along these lines are wedge issues historically leading to sharp disagreements between anthropocentric, ecocentric, and biocentric theorists and practitioners.

Species/Ecosystems Impacts

If species and ecosystems are included in an assessment, these biotic entity types might be narrowed to imperiled species and/or ecosystems within the geographical scope of the policy or project evaluated. As a basis for selecting predicted impacts that have a significant bearing on endangered species and/or ecosystems, a suggestion approach involves an initial identification of key indicators used in conservation biology to assess species and ecosystem decline and health (e.g., loss or reappearance of a keystone predator, extensive loss or reestablishment of critical habitat, intrusion or elimination of invasive species, and the like). Having identified relevant key indicators, two lists of predicted impacts would be compiled: one list that includes impacts expected to contribute in a significant way to species and/or ecosystem recovery; the other list including impacts expected to significantly contribute to their continuing decline. Predicted impacts in each of the two sets should not involve an indiscriminate re-listing of all of the predictions pertaining to each of the other biotic entity types. Prior identification of key indicators serve as criteria for selecting what predicted impacts are particularly pertinent in regard to species and/or ecosystem decline or recovery.

As with value weighting, scalers would then be used to estimate the degree to which each previously predicted impact (included on one of the two lists) contributes to or diminishes the prospect of species and/or environmental enhancement or decline, where scalers ranging between 1 and 3 are used to rate positive contribution and scalers ranging between -1 and -3 are used to rate negative impact. These assignments (perhaps arrived at though consultation with experts) would then be added or subtracted as the case may be to the initially computed impact magnitudes for entity type "imperiled species and/or ecosystems."

Model Uses

Numerous research organizations in government agencies, universities, and nongovernmental organizations (NGOs) have the organizational and personnel resources to conduct an assessment along the lines proposed, and it is not hard to imagine an entire impact assessment organized and conducted online, perhaps through a website set up for this purpose. The model and the suggested approach for developing and validating inputs could be used to evaluate a wide variety of proposed policies or projects at local, national, or international levels in both the public and private sectors.

The model facilitates "what if" analysis by translating scenarios into model input values and comparing recomputed indices at the aggregate entity and/or project level with those of the baseline. It could also be used to compare a proposed policy or project with alternative courses of action.

While the model might be applied in some of these ways, the underlying moral claim is that policy makers and concerned citizens should ask the questions that are built into the model's structure. They ought to ask whether and how nonhuman life and life systems could be expected to be harmed by policy and project proposals, and they ought not act on decisions whose benefits are not greater than its hams, all potentially impacted biotic entities taken into consideration.

Assessment Management

Outside a classroom learning environment, a rigorous assessment along the lines suggested will require a qualified assessment manager with broad knowledge across multiple scientific disciplines, and also the organizational and leadership skills needed to ensure that participants in the assessment work as a cohesive and coordinated team. Key management tasks include: (1) structuring the assessment by making flexible front-end decisions on the appropriate level of aggregation of biotic entity types, harm and benefit types, and impact degree-of-magnitude parameters based on the information needs of decision makers and usefulness for impact prediction; (2) development of a detailed description of the policy or project to be evaluated, including its justificatory rationale, an implementation and operations plan, a plan for dealing with contingencies and unforeseen circumstances, a timeline for all phases of policy/project implementation, and rough cost estimates; (3) assembly of a team to perform defined responsibilities for making, gathering, and organizing impact predictions and for estimating probabilities and parameter estimates as part of an iterative, multi-round process involving previous round summaries and next-round feedback (which process will require the use of forms and questionnaires for eliciting information and preparing feedback reports); and (4) ensuring that final-round results are documented and made available to interested parties for review, challenge, and further refinement.

APPENDIX: DRILLING IN THE ALASKA NATIVE WILDLIFE REFUGE (CASE STUDY)

A Sense of the Refuge

The Alaskan Native Wildlife Refuge (ANWR) is in northeast Alaska and today consists of about 19.3 million acres, a land area approximating that of South Carolina. Established in 1960 as the Arctic National Wildlife Range, it was renamed the Alaskan Native Wildlife Refuge with the passage in 1971 of the Alaska National Interest Lands Conservation Act. The Refuge has no roads, marked trails, or campgrounds and is managed by the US Fish and Wildlife Service with the mandate to preserve unique wildlife, wilderness, and recreational values. The Refuge comprises five ecological regions:

- The north coastal marine region, next to the Arctic Ocean, consists of salt marshes, lagoons, barrier islands, beaches, and river deltas. These areas are home to polar bears, fish, and migratory birds.
- The coastal plain region lying to the south is a treeless, flat-to-hilly region where caribou and birds raise young in summer. Polar bears den in the coastal plain in winter and hunt fish in deep pools along shallow rivers after hibernation.
- The alpine zone consists of the vast Brooks Range, the northern-most extension of the Rocky Mountains. Dall sheep, grizzly bears, wolves, and ground squirrels live there.
- The forest-tundra transition has spruce trees interspersed with low tundra plants. Caribou feed in this area through the winter, and moose and wolves roam there year-round.
- The southernmost boreal forest region is a mix of spruce, birch, and aspen trees and is the only extensively forested area in the Refuge. It is the home to moose, lynx, weasels, and numerous songbirds.

Native peoples have lived in the northern and southern portions of the area for thousands of years. Inupiat Eskimos live along the northern coast. They hunt seals and whales, fish from barrier islands and along freshwater rivers, catch birds throughout the spring and summer, and travel inland to hunt caribou, wolves, and Dall sheep. Gwinch'in Athabaskan Indians live to the south. Traditionally, they travel widely each year in search of food and still establish summer fishing camps along rivers, catch waterfowl and other birds, snare small game, and hunt caribou.

Summer in the Arctic is short and the days are long. On the coastal plain, the sun circles the horizon twenty-four hours a day and does not set from mid-May until late July. For those two intense months, midday and midnight sun blend into one, and plants, animals, and people work hard to garner the resources needed for the long winter season ahead. The winter moon, as the summer sun, circles overhead without setting. Reflections off the expanse of snow illuminate the long nights, and the aurora borealis (the "northern lights") periodically glows in the starry sky.

Plants and animals have adapted to these harsh conditions in a variety of ways. A polar bear mother spends five months in her winter den. Mosquitoes survive winter by replacing the water in their bodies with substances that act like antifreeze in a car. Muskoxen save energy by moving very little and are protected by foot-thick coats. Similar to plants, lemmings use the blanket of snow as insulation. The base of a snow pack may be +20°F, while the air at its surface may be -30°F. The Arctic fox's short legs and small ears help it retain body heat. Of the Refuge's two hundred recorded bird species, all but about twenty-five migrate to warmer regions. Some (e.g., the Arctic tern) make round trips up to twenty-five thousand miles. Caribou stay active all winter and exhibit the longest migration of any land mammal, up to 2,500 miles annually, traveling in herds between winter grazing ground in the south and summer calving grounds in the north. Plants are the foundation of the food chain in the Arctic Refuge. They are eaten by grazers and the latter are eaten by predators.

Warming as much as five to seven degrees Fahrenheit has occurred over the past fifty years. If current trends continue all the glaciers in the Arctic Refuge may disappear by the end of the century. Defenders of Wildlife report that ten Arctic Refuge species—lynx, wolverine, caribou, Dall sheep, Alaska marmot, arctic ground squirrel, singing vole, northern bog lemming, tundra shrew, and barren ground shrew—are "highly vulnerable" to climate change.

Three mammals—the brown bear, marten, and yaiga vole—are classified as "moderately vulnerable." Species adapted to cold, snow, and ice are particularly vulnerable from snow to freezing rain shifts and the collapse of snow tunnels animals use to reach plant nutrients on the ground. To protect these vulnerable species, environmentalists cite the need to limit activities that disturb wildlife and destroy habitat, to maintain linkages to tundra adjacent to the Refuge, and to accelerate investment in research and monitoring of vulnerable species.

Water is the lifeblood of ANWR. It stings as wind-driven snow in the winter, soaks as gentle rain in summer, and penetrates as thick fog. Fed by springs and river flow during winter, it piles up in fields of ice. In spring, it floods the river corridors, washing gravel bars and nourishing plants. During summer, it courses through rivers and streams and finally reaches the north coast, enriching the ocean with its load of minerals. All along the way, it provides life-sustaining habitats for invertebrates and fish, nesting for birds, and drinking water for animals and humans. Water quantity is limited, especially on the coastal plain, where less than five inches of precipitation falls each year. The plain itself has few lakes, and they are shallow and unevenly distributed. Most of the water available during summer comes from spring snowmelt. It pools on the surface and little percolates through the foot-thick permafrost. Water rights in the area are federally owned and managed by the US Fish and Wildlife Service in cooperation with the state.

When the Refuge was established in 1960, it was already receiving occasional visits by hunters and hikers. Visitation increased gradually throughout the 1970s and '80s. A public use study in 1975 estimated 281 visitors. By 1986, that figure doubled and the number of commercial recreational guides increased from seven in 1980 to 21 nine years later. Since 1986, visitor numbers have more than tripled. People generally stay seven to eleven days, most often to backpack, float rivers, and hunt.

Legal and Political Context

The following summarizes key events related to the establishment of the Alaska Native Wildlife Refuge and the long-contested debate over whether oil and gas drilling should be allowed in this pristine region.

1959 Signing of the Statehood Act by President Kennedy, giving the state of Alaska the right to choose 103 million acres for use as an economic and tax base.

1964 Signing of the Wilderness Act by President Lyndon Johnson, establishing the National Wilderness Preservation System.

1967 Protest of a federal North Shore oil and gas lease by Alaska Natives, resulting in a freeze on all federal lands in Alaska pending settlement of Native land claims.

1971 Signing of the Alaska Native Claims Settlement Act, giving the Kaktovik Inupiat Corporation surface rights to 92,160 acres of federal lands adjacent to the village and 69,120 acres to be selected within the Arctic Range. The coastal village of Kaktovik is the only settlement in the Arctic Refuge.

1977 Completion of the Trans-Alaska Pipeline System bringing North Shore oil to the mainland.

1980 Signing of the Alaska National Interest Lands Conservation Act (ANILCA) by President Richard Nixon, creating 104 million acres of national parks, wildlife refuges, and wilderness areas in Alaska. The act expanded the Arctic Refuge to approximately eighteen million acres, designated eight million acres as wilderness, and designated three rivers as wild. Section 1002 of ANILCA forbids drilling on the coastal plain (containing the Section 1002 area) without congressional approval.

1983 Expansion of the Refuge by nearly one million acres in the south when the state decided not to retain control of lands it had selected under the Statehood Act.

1983 Conveyance per the Chandler Lake Land Exchange Agreement of federal subsurface rights on Kaktovik Inupiat Corporation (KIC) lands to the Arctic Slope Regional Corporation, allowing exploratory drilling on KIC lands subject to congressional approval.

1986 Recommendation by the Republican-led US Fish and Wildlife Service (in a draft management plan) that the 1002 area be opened for oil and gas development to protect the nation's economy and national security. In response, Democratic Congressman Morris Udall, chairman of the House Interior Committee, unsuccessfully introduced legislation to turn the 1002 area into a protected wilderness area.

1987 Signing of an agreement by the US and Canada to protect the porcupine caribou from damage to its habitat and migration routes. Canada's Ivvavik and Vuntut national parks border ANWR. The agreement requires consultation prior to any final decisions.

1988 Addition of 325,000 acres to the south end of the Refuge, bringing the total area to 19.3 million acres and making it the largest refuge in the National Wildlife Refuge System.

1989 Exxon Valdez oil spill, effectively derailing a pending bill that would have authorized 1002 drilling.

1990–95 Repeated initiatives in Congress related to 1002 drilling. Various pro-drilling strategies were employed, including threatened filibusters and attempts to add provisions to the Energy bill, the Defense Appropriations bill, and the budget resolution.

1996 Veto by President Bill Clinton of a bill passed by the Republican-controlled Congress to allow 1002 drilling.

1997 Ruling by the US Supreme Court that the barrier islands and lagoons that they enclose are within the boundaries of the Arctic Refuge.

2012 Release in August 2012 by the US Fish and Wildlife Service (USFWS) of its Revised Comprehensive Conservation Plan for ANWR. The plan outlined six management alternatives: (A) no action; (B) designation of the Brooks Range as a wilderness study area; (C) designation of the Coastal Plain (the 1002 area) as a wildlife sanctuary; (D) designation of the Brooks Range and the Porcupine Plateau as wilderness study areas; (E) designation of the Brooks Rane, the Coastal Plain, and the Porcupine Plateau as wilderness study areas; and (F) no new wilderness study areas with revised management guidelines.

2015 Formal adoption of Alternative E under the USFWS Comprehensive Management Plan for ANWR, under which the designated lands will be managed under the Minimal Wilderness and Wild River Management categories described in EIS Chapter 2, Section 2.3, for the next fifteen years.

2016	Introduction of an amendment of House Bill 961 declaring Section 1002 of the Coastal Plain area as "wilderness" as defined in the 1964 Wilderness Act. The bill was defeated by fifty-one votes in a roll call vote ending at 176–277.
2017	President Donald Trump signed into law P.L 115-97 ("Tax Cuts and Job Act"), which authorizes the secretary of the Interior to administer an oil and gas leasing program in Section 1002 subject to the completion and approval of an environmental impact statement.

Estimates of ANWR Oil Reserves

In 1998, the US Geological Survey (USGS) estimated that between 5.7 and 16.0 billion barrels of oil are technically recoverable from ANWR's coastal plain, of which a mean estimate of 7.7 million barrels falls within the 1002 area. The report estimated that 120 billion barrels are technically recoverable from other areas in the US.

A report prepared in 2008 for Alaska Senator Ted Stevens by DOE's Energy Information Administration, and based primarily on the 1998 USGS report and the National Energy Modeling System, came to the following conclusions:

- Assuming no regulatory delays, ANWR oil production would begin ten years after legislation approves oil and gas leasing in the 1002 area, during which time an environmental impact statement would be approved and leases acquired (2–3 years), a single exploratory well would be drilled (2–3 years), a production development plan would be prepared (1–2 years), and feeder pipelines, oil separation and treatment plants, drilling pads, and wells would be constructed (3–4 years).
- If production began in 2018 (2000 + 18 years), production of 780,000 barrels per day would reach a peak in 2027 and decline to 710,000 barrels in 2030 (mean case).
- ANWR production would represent a small fraction of total world production, and would very likely be offset, in part, by somewhat lower production outside the US. [DOE estimates proved (vs. prospective) world reserves in the range of 1.1–1.3 trillion barrels.]

- ANWR production would reduce oil import dependency to 47.5 percent (mean case) during the 2022–2026 time frame, but with declining production would increase again to 54 percent by 2030.
- The opening of ANWR is projected to reduce the price of light crude oil by $0.75 per barrel (2006 dollars).
- The upper limit to ANWR oil production is the transportation capacity of the Trans-Alaska Pipeline System (TAPS). The high ANWR oil production case is projected to come closest in 2026 to reaching TAPS' pipeline capacity (2.136 barrels per day), when total North Slope oil production peaks at 1.9 million barrels per day.

In 2010, USGS downsized estimates of recoverable oil from the National Petroleum Reserve (east of Prudhoe Bay and west of ANWR) following its conclusion that areas previously believed to hold oil were primarily gas producers.

Views Opposing 1002 Drilling

- The 2008 DOE report cited previously concluded that "ANWR production is not projected to have a large impact on world oil prices. ... Additional oil production resulting from the opening of ANWR would be only a small portion of total world production, and would likely be offset by production outside the United States."
- The Natural Resources Defense Council (NRDC) contends that 1002 drilling would not take place in a compact two thousand-acre space as proponents claim, but require "a spider web of industrial sprawl across the whole of the refuge's coastal plain, including drill sites, airports, roads, and gravel mines. Its actual footprint would be spread across 640,000 acres, or 1,000 square miles."
- USGS reports that the 1002 area "has a greater degree of ecological diversity than any other similar sized area of Alaska's north slope," and that "those who have campaigned to establish the Arctic Refuge recognized its wild qualities and [its] unusually diverse assemblage of large animals and smaller, less-appreciated life forms, tied to their physical environments and to each other by natural, undisturbed ecological and evolutionary process."

- Numerous Native American groups oppose oil and gas development within the Refuge. The Alaska Inter-Tribal Council (representing 229 Native Alaskan tribes) opposes such development. The Native Alaskan and Canadian Gwinch'in Tribe views ANWR drilling as "a basic Aboriginal human rights issue," noting that "Sixty to 70 percent of our diet comes from the land and caribou is one of the primary animals that we depend on for sustenance." The Inupiat from Point Hope, Alaska, passed a resolution in 2005 expressing concern that 1002 drilling would lead to resource exploitation in other wilderness areas. The National Congress of American Indians and the Native American Rights Fund also oppose drilling in the Refuge.
- Organizations voicing strong opposition to 1002 development during the public comment period following release of the USFWS 2012 draft Revised Comprehensive Conservation Plan included Defenders of Wildlife, Greenpeace, Sierra Club, Natural Resources Defense Council, Center for Biological Diversity, The Wilderness Society, Alaska Wilderness League, Save our Environment Action Center, National Wildlife Federation Action Fund, League of Conservation Voters, Audubon Society, Endangered Species Coalition, Wilderness Watch, Pacific Environment, National Wildlife Refuge Association, Gwinch'in Nation, Republicans for Environmental Protection, and Friends of Alaska National Wildlife Refuges.
- In response to a League of Conservation Voters survey conducted during the 2008 presidential campaign, candidate Barack Obama said, "I strongly reject drilling in the Arctic National Wildlife Refuge because it would irreversibly damage a protected national wildlife refuge without creating sufficient oil supplies to meaningfully affect the global market price or have a discernible impact on US energy security." John McCain, Obama's rival on the Republican ticket, expressed this sentiment: "As far as ANWR is concerned, I don't want to drill in the Grand Canyon, and I don't want to drill in the Everglades. This is one of the most pristine and beautiful parts of the world."

Views Supporting 1022 Drilling

- A majority of Alaskans support development, including every governor, senator, representative, and legislature for the past twenty-five years.

Alaska residents receive annual dividends ($2,072 per resident in 2015) from a permanent fund dependent primarily on oil and gas revenues.

- The Resource Development Council for Alaska, an association of companies from the oil and gas, mining, forest products, tourism, and fisheries industries, has consistently supported "the responsible development of the State's natural resources."
- In its active and ongoing campaign for ANWR development, Arctic Power points out that precedents exist for drilling in wildlife refuges, citing a US Fish and Wildlife Service report that 77 of the 567 wildlife refuges in twenty-two states permit oil and gas-related activities, with actual production occurring in forty-five refuges in fifteen states.
- A state-funded survey revealed that fifty-three of sixty-eight responding Kaktovik residents "agree that ANWR's coastal plain should be opened to oil and gas exploration." While the survey instrument and tabulation of results is alleged to be problematic, it is evident that some support among Kaktovik residents exists for further exploratory drilling.
- In response to the USFWS 2012 draft Revised Comprehensive Conservation Plan, the Office of the Governor said, "Oil and gas development and production in the Refuge may be authorized by Congress at any time, and the current national dialogue regarding the need for jobs, energy security, and deficit reduction makes the likelihood of such an action higher than ever before."

North Sea Oil Exploration and Development

Following federal approval of an air quality permit and an oil spill cleanup plan, Royal Dutch Shell has embarked on an exploratory drilling program to assess whether an estimated twenty-seven billion barrels of oil and 132 trillion cubic feet of natural gas is recoverable from Alaska's North Sea. Proponents argue that production could cut America's dependence on foreign oil and provide jobs and needed revenue to the Treasury Department. Opponents fear that drilling rig and tanker spills could lead to unmitigated disaster.

The Center for Biological Diversity warns that "All signs point to environmental disaster if an oil spill were to occur in the harsh Arctic waters." The Alaska Wilderness League agrees: "The Arctic is prone to hurricane-force storms, 20-foot ice swells, sea ice up to 25 feet thick, sub-zero temperatures, and months-long darkness. What's more, the Arctic has extremely limited infrastructure (there are no roads or deep water ports and only a handful of small airports) and the nearest Coast Guard station is 1,000 miles away." The Natural Resources Defense Council adds that "Nobody knows how to contain or clean up a spill in the harsh remote seas of the Arctic. If a blowout occurs late in the drilling season, there won't be time to stop it before the winter ice chokes the site, leaving oil to gush uncontrolled for months." Perhaps mitigating this concern, scientists predict there will be less ice by mid-century, and some are predicting that year-round shipping will be common by that time.

In Washington, politicians are jockeying over whether to ratify the "Law of the Sea Treaty," under which countries abutting oceans will be able to claim up to two hundred miles of undersea territory if they can prove it to be an extension of their continental shelves. For the US, that could mean extra territory the size of California. President Obama and former President George W. Bush support the treaty, as does an oddly aligned coalition that includes the Pentagon, oil companies, shipping companies, and various environmental organizations that favor the part of the treaty designed to help protect the world's oceans. Although every other Arctic country has ratified the treaty, conservative senators fear that ratification would give too much influence to multinational bodies.

Shell Oil began exploratory drilling in in two areas of the North Sea in 2012: a site west of Barrow named Burger in the Chukchi Sea, and a site close to Kaktovik named Hammerhead. Efforts to date have not gone well for Shell. Drilling had to be postponed from July to August of 2012 because of an unexpectedly thick ice pack. Also, concerns have been raised about the durability of one of Shell's underwater oil spill containment vessels; numerous small spills have occurred, and two of its drilling rigs, the Kulluck and Noble Discoverer, are on their way to Asia for dry dock repairs. En route for repair, the Noble Discoverer slipped its mooring and ran aground near one of Alaska's Aleutian Islands, whereupon Shell announced suspension of drilling activities. In 2017, Shell announced that it plans to resume exploratory drilling in the Chukchi Sea in the summer of 2019.

Meanwhile, Norway and Russia have moved to develop their Arctic resources. Norway's northmost drilling platform is scheduled to begin production in the summer of 2019, with plans to open more areas for drilling in the Barents Sea. Russia has a producing field in the Pechora Sea (the target of Greenpeace protests in 2012 and 2013), and plans four more wells on this field in 2019. Russia is also moving forward on major projects on the Yamal Peninsula, including a deepwater port and an LNG production plant to export Yamal's vast gas reserves to Europe or Asia.

Not surprisingly, Alaska's indigenous residents view North Sea drilling as ominous, if not disastrous. The village of Kaktovik passed a resolution in May 2006 authorizing the mayor to take legal and other actions necessary to "defend the community." The resolution referred to Shell Oil as "a hostile and dangerous force" and "called on all North Slope communities to oppose Shell offshore leases … until the company becomes more respectful of the people." Kaktovik's mayor stated that Shell has failed to work with the villagers on how the company would protect bowhead whales, which are part of Native culture, subsistence, and diet.

Concluding Note

The future of oil development in the region depends on the outcome of Interior's environmental impact review of the 1002 leasing program authorized by P.L. 115-97 and legal challenges. As the CEO of Defenders of Wildlife remarked following passage of this legislation, "We've been fighting to protect the Arctic National Wildlife Refuge for more than thirty years. We won't let it end like this." The future of oil and gas development in ANWR also depends on whether exploratory drilling in the North Sea reveals economically feasible recoverable reserves, whether the leasing program authorized by P.L. 115-97 proceeds, and, ultimately, whether Congress authorizes production as required by the 1980 Alaska National Interest Lands Conservation Act (ANILCA).

INDEX

D

E

F

G

S

T

www.ingramcontent.com/pod-product-compliance
Lightning Source LLC
LaVergne TN
LVHW080652011026
849160LV00014B/1395